SpringerBriefs in Public Health

For further volumes:
http://www.springer.com/series/10138

Niraj Kumar

Biogeogens and Human Health

Niraj Kumar
Science and Society Division
Institute of Applied Sciences (IASc)
Allahabad
Uttar Pradesh
India

ISSN 2192-3698 ISSN 2192-3701 (electronic)
ISBN 978-81-322-1083-2 ISBN 978-81-322-1084-9 (eBook)
DOI 10.1007/978-81-322-1084-9
Springer New Delhi Heidelberg New York Dordrecht London

Library of Congress Control Number: 2013938753

Printed on acid-free paper

Springer is part of Springer Science+Business Media (www.springer.com)

Foreword

The sustenance and renewal of life in an ecosystem depends mainly on two factors, energy cycling and nutrient flow. Nitrogen cycle and nitrogen fixation are perhaps the best examples of nutrient cycling. Nutrient cycling involves elements moving to/from biotic and abiotic factors in the ecosystem. Energy movement through food chain in biotic factors is an important example of energy flow. Energy is either transformed or conserved.

Plants and some other organisms convert solar energy to chemical energy via photosynthesis which occurs in two parts: light-dependent reactions and dark reactions. The light-dependent reaction happens when solar energy is captured to make adenosine triphosphate (ATP). The dark reaction happens when the ATP is used to make glucose (the Calvin Cycle). Chlorophyll on irradiation with sun light forms antenna complexes which transfer light energy to one of two types of photochemical reaction centers: P_{700}, which is part of Photo-system I, or P_{680}, which is part of Photo-system II. Excited electrons are transferred to electron acceptors, leaving the reaction center in an oxidised state.

The role played by an organism in its environment is defined as "Niche" in Ecological Biology. A niche may also encompass how the organism interacts with other living things or biotic factors, and also with the non-living, or abiotic parts of the environment as well. All living organisms have a fundamental niche. This is all of the possibilities available for the organism to take advantage of. All possible sources of food, all open roles in the environment, and any suitable habitat is included in a fundamental niche. Any organism does not use all the available resources at the same time. Rather it uses in a very narrow range. This more specific role is called the organism's realised niche. Abiotic factors, such as water availability, climate, weather, and in the case of plants, soil types, and amount of sunlight can also narrow a fundamental niche to a realised niche. An organism can somewhat adapt to its environment, but the basic needs must be met first in order for them to have time to find their niche.

Abiotic disorders can influence biotic disorders and vice versa. Biotic and abiotic factors are closely linked in an ecosystem. The environment affects the human health in a big way. People tend to be most susceptible to illness when physically or mentally stressed. Stress, energy and immunity form a closely knit network.

The author in the present book has brought out this intricate concept of interdependence of biotic (living) and abiotic (non-living) factors in an ecosystem, resulting in an impact on human health, in an explicitly marvelous manner. As a result a new word "Biogeogens" has been coined, "bio" for living (biotic), "geo" for non-living (abiotic/geographical/climatic/environment) and "gens" for the interactive proceeds of the two. For the ambience embedded with clarity with which the author has explained the flow of energy cycle through these two important factors, "biotic" and "abiotic" and how it influences human health is praiseworthy. I congratulate the author for this Herculean task and hope that readers will find the book useful and interesting.

Prof. Krishna Misra, FNASc
Coordinator, Indo-Russian Centre for Biotechnology
Indian Institute of Information Technology, Allahabad
NASI-Senior Scientist, CBMR-SGPGI, Lucknow
General Secretary, The National Academy of Sciences, India

Preface

The WHO data on the burden of diseases suggest that approximately 80 % health problems in rural parts of the India are due to communicable diseases like diarrhea, typhoid, cholera and infective hepatitis etc. Only diarrhea kills approx. 6 lakhs children in India every year. Morbidity and mortality due to malaria are the major public health concerns with around two million cases reported annually; Filaria, T.B., Measles, Flu and worm infestations add further burden. The 50 % of health budget is spent in tackling health impact of disease related to water pollution as vector of communicable diseases in India. Therefore, preventive and promotive health care mechanisms are must to improve the situation.

Realising this very fact, the Institute of Applied Sciences, Allahabad, India, actively involved in the application of scientific know how for the betterment of society has undertaken many research projects to evaluate the geography of health of suburban and rural population in and around Allahabad. The guiding principles to conduct these projects were mainly inspired by the studies of Prof. Rais Akhtar, a noted medical geographer and an eminent scientist of this country. But, the author further analysed several other factors affecting human health, based on his participatory observations done to evaluate the nutritional requirements, water intake, food habits and associated taboos, disease condition and health status of the population under study. Finally, on the basis of the years of strenuous and methodological studies done under the guidance of the author and involving the researchers Prasanna Ghosh and Vartika, a Dogma of Biogeogens has been postulated by the author—which could be helpful for the future scope of research in this field as well as to cast out many doubts in effective management of human health.

From Author's Pen

Acknowledgments

I deem it to be my proud privilege to take this opportunity to pay my deepest regards and gratitude to the past Presidents of the National Academy of Sciences, India (Hon'ble Prof. A. K. Sharma and Prof. Manju Sharma) and the present President (Hon'ble Dr. K. Kasturirangan) of the Academy; and the Institute of Applied Sciences, Allahabad (Dr. B. P. Agrawal), for their continuous support, inspiration and encouragement that has made this work possible.

I also acknowledge a deep sense of gratitude to Hon'ble Prof. V. P. Sharma (Past President, NASI) and Prof. Krishna Misra (General Secretary, NASI), whose noble guidance, valuable suggestions and advice always helped me to move in a right direction, and do something meaningful to the service of science and society.

It is my genuine abiding desire and great pleasure to express my most sincere thanks to my colleagues Dr. K. P. Singh, Vice-President of IASc, Allahabad, and Dr. Ashwani Kumar, IFS, Former Director, Forest Research Institute, U.P., for their kind support throughout the period of the work.

I express my foremost thanks to my beloved researchers—Prasanna Ghosh, Vartika and Fatmatuz Zohra, for their keen interest in the research projects and carrying out the studies with full devotion and dedication, due to which I could prove my hypothesis.

In the end, my sincere thanks to Dr. Mamta Kapila and Sri Aninda; and all my love, affection and good wishes to my heartiest friends and family members for their helpful and cooperative nature and attitude, especially to my lovely better half and three kids—Aashu, Apaarna and Twinkle.

With all my best regards to my parents (Mother—Smt. Dharmashila Srivastava and Father—Late Shri K. P. Srivastava).

From Author's Heart

Contents

Biogeogens and Human Health

Introducing the Concept of Biogeogens

The world mainly depends on the transformation and conservation of energy from one form to the other. Energy is transformed and conserved in terms of its channelisation in the biotic and abiotic resources, which could be explained by certain phenomena based on the principle of $E = mc^2$.

Energy is neither created nor destroyed. Meaning simply that energy is either transformed or conserved. It is also defined that energy is the ability to do work. This cause and effect relation is directly linked with all the biotic and abiotic factors and phenomenon. They could be made understandable in terms of energy, its flow, loss, transformation and conservation. A living system gets energy in the chemical form; some machines also get their energy in chemical form, i.e. from fuels like gasoline or diesel. A plant has chemical energy stored in its leaves, which happens due to a complex phenomenon/process, i.e., photosynthesis.

Photosynthesis is the process that supports life on earth. Plants use the sun's energy to transform CO_2 and water into a sugar called glucose.

$$6CO_2 + 12H_2O = C_6H_{12}O_6 + 6H_2O + 6O_2$$

The process of photosynthesis starts with an advent of light on the leaves of plants. The sun's energy is transferred to plants by means of photons hitting the chlorophyll.

N. Kumar, *Biogeogens and Human Health*, SpringerBriefs in Public Health,
DOI: 10.1007/978-81-322-1084-9_1,

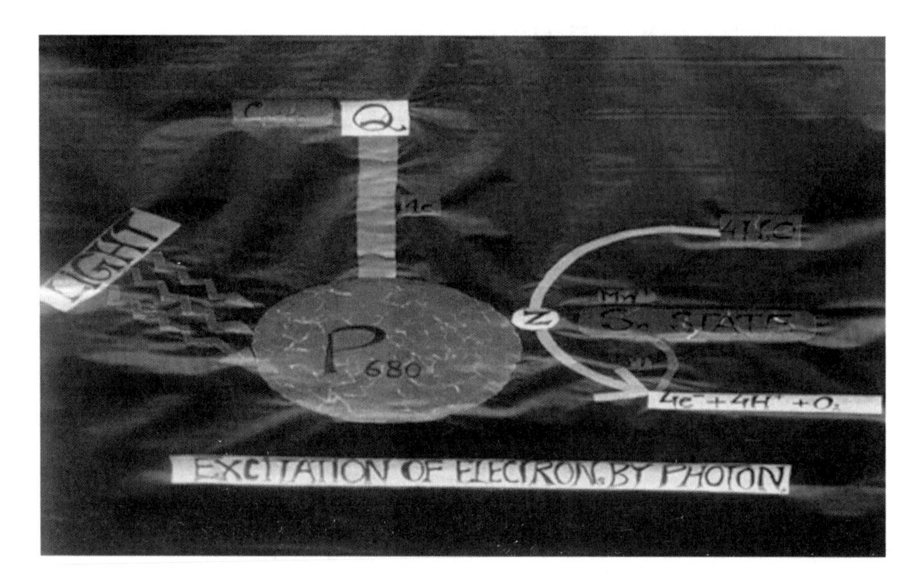

Diagrammatic representation of electron release on incident of light on the chlorophyll [chart made by Ms Fatmatuz Zohra, Vigyan Sancharak Fellow (VSF), DST, New Delhi under the mentorship of the author in Institute of Applied Sciences (IASc), Allahabad]

This stored energy can be released if the chemical makeup of the plant changes, for example if it is eaten by an animal/human being or burn to produce heat and light. That is how the energy stored in the form of chemical/food energy is transferred to other living organisms, i.e., from producers to consumers through different trophic levels; in each step, a sufficient amount of energy is consumed in other processes before transfer. This consumption/transfer of energy from one process/system to another involves many phenomena and factors, which interact with each other in an orderly manner leaving an impact on many processes of life, thus:

- The living and nonliving are interwoven on the matrix of energy dynamics, in terms of either electron flow or photon flow.
- The cycle of food chain starts with the transfer of energy from photons to electrons and goes on to the extent of conservation/transformation of energy from one form to other (e.g., from food/chemical energy to heat energy and so on).
- Further availability of this food energy decides the fate of an ecosystem, correlating the phenomenon of biogeochemical cycle with the food web.
- Finally, the pray–predator and its population dynamics play a pivotal role in deciding the pyramid of energy, (as seen in the photograph- courtesy: VSF, IASc, Allahabad).

Thus biotic and abiotic factors interact with each other; and selection pressure decides the directional flow of the energy-economic dynamics. All this creates a multidimensional impact on the biotic and abiotic systems. That is how the two components are linked together in an ecosystem.

As per the United Nations Environmental Programme (1992) for planet protection, "All constituents of the environment of our planet ultimately exert an influence on human health and well-being". Further, if one has to define all the constituents (abiotic and biotic) in an environment in terms of their relation to each other, as described above, and their impact on human health, biotic may be abbreviated as *bio* while abiotic (non-living/geographical/climatic) as *geo*; and *all those proceeds which are supposed to arise from the processes taking place in these two* may be termed together as *Biogeogens*. Now, the question is what are these in real terms, which could affect human health and how?

Before discussing these factors, it is necessary to define first human health; only then would it be possible to understand the effect of biogeogens on it. *Human health is a state of complete physical, mental and social well-being and not merely the absence of disease or infirmity* (as defined by World Health Organisation). The essential requisites of "human health" include the following:

- achievement of optimal growth and development as per genetic make-up,
- maintenance of the structural and functional efficiency of body organs/tissues necessary for an active and productive life,
- mental alertness and well-being,
- ability to withstand the inevitable process of ageing with minimal deformity/ disability and functional impairment, and
- ability to tolerate and combat diseases.

Till about three decades ago, the importance of nutrition as a major determinant of health had been widely recognised only with respect to the first two of the requisites listed above. But now, it is a well-established fact that nutrition plays a major role in negating the effects of toxins and pollutants [1]. One major pre-requisite for the maintenance of health is that there be an optimal dietary intake of a number of nutrients; the chief of these are vitamins, certain amino acids, certain fatty acids, various minerals, and water [2, 3]. Each of these aspects has a deep influence on health and which in turn influences all these aspects.

Thus, nutrition is a major process taking place in the human body giving rise to several free radicals, antioxidants, etc.; all these could be considered as *biogens* (*the proceeds of the processes taking place in a biotic component*), which interact with the *geogens* (radiations, pesticide and other chemicals, etc., i.e., *the proceeds of the processes taking place in an abiotic component*), determining finally the status of human health.

From the above, one can understand simply the connotation of biogeogens; however, assessment of its impact on human health is not easy as the adaptive interactions of biogeogens further give rise to several other complications making the dynamics more and more complex, necessary to be deciphered for regulating the human health, which will be discussed in detail to fix-up their role in determining the well-being of humans.

References

1. Bamji SM, Rao PN, Reddy V (2003) Text book of human nutrition: in the expanding frontiers of nutrition science, 2nd edn. Oxford and IBH publishing Co. Pvt. Ltd., New Delhi, p 1–20
2. Kornberg A (1992) A basic research: the life line of medicine. FASEBJ 6:3143
3. Robert MK, Daryl GK, Mayes PA et al (2001) A lange medical book: biochemistry and medicine, Harper's Biochemistry, 25th edn. p 1–10

Dogma of Biogeogens

Biogeogens arise together as an aftermath of interplay of the proceeds of the processes taking place under the influence of the following dimensions:

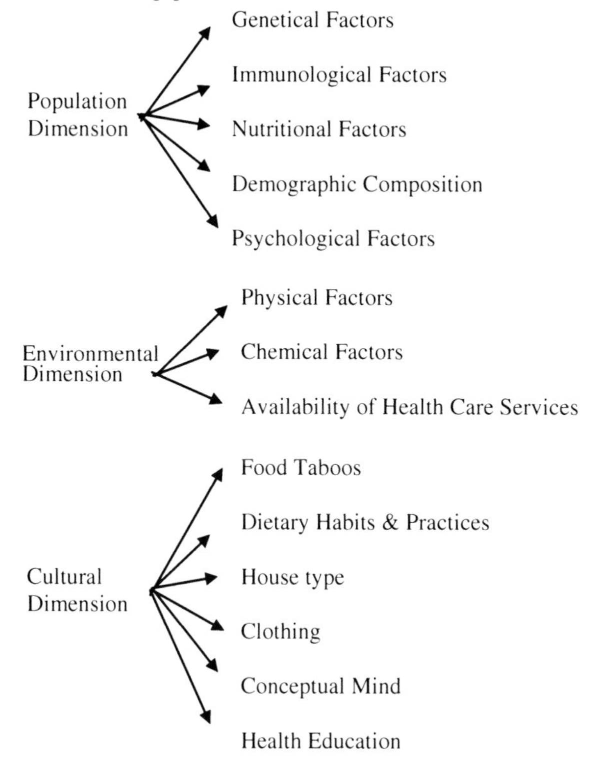

Each and every dimension has its own significance, and is further categorised into several components:

N. Kumar, *Biogeogens and Human Health*, SpringerBriefs in Public Health,
DOI: 10.1007/978-81-322-1084-9_2, © The Author(s) 2013

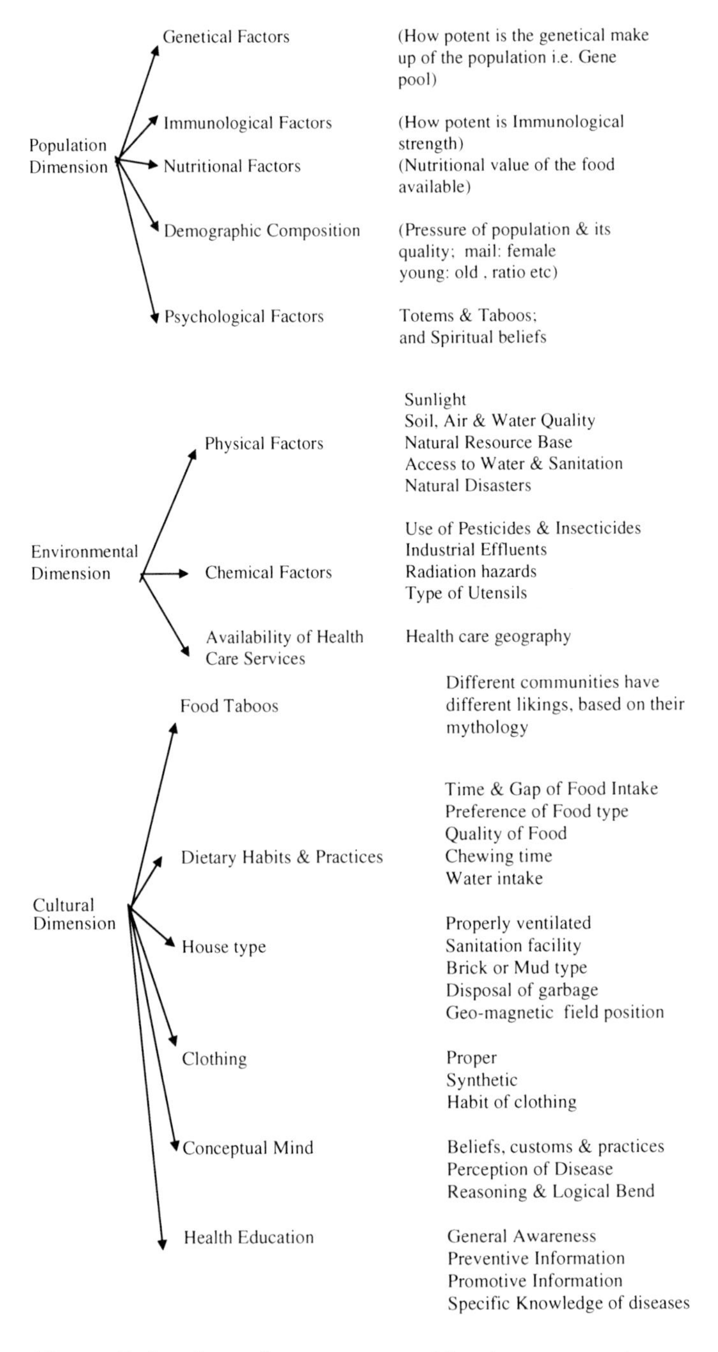

Thus, all the above factors are considered necessary to assess the status of health; the health problems of any community are influenced by an interplay of all these factors [1]. The common beliefs, customs and practices related to health and disease in turn influence the health-seeking behaviour of the community [2]. It has been shown in different studies that various socioeconomic, mental and other factors affect the nutritional status of infants/children [3–5]. In other studies also, it has been well established that the nutritional status of a person is a function of his/her socioeconomic condition [6–9].

Further, widespread poverty, illiteracy, malnutrition, absence of safe drinking water and sanitary living conditions, poor maternal and child health care services and ineffective coverage of national health and nutritional programmes/services have been traced out in several studies as possible contributing factors to dismal health conditions prevailing among the rural/tribal population in India [10–21]. Hence, a comprehensive approach in the light of the concept of biogeogens is needed to reach any definite conclusion regarding the health status of an individual/community.

References

1. ICMR Bulletin (2003) Health status of primitive tribes of Orissa. ICMR Bull 33 (10):9
2. Indian Council of Medical Research (1998) Health of Tribal population in India; result of some ICMR studies. Indian Council of Medical Research, New Delhi
3. Indian Council of Medical Research (1977) Studies on preschool children. ICMR Technical report series: 26, New Delhi
4. Chaudhury RL (1984) Determinants of dietary intake and dietary adequacy for pre-school children in Bangladesh. Food Nutr Bull 6(4):24–33
5. Tara G, Poonal P, Meenakshi B (1988) Women and nutrition: selected socio-economic, environmental, maternal and child factors associated with the nutritional status of infants and toddlers. Food Nutr Bull 10(4) (The United Nations University Press)
6. Bharati P, Basu A (1988) Uncertainties in food supply and nutritional deficiencies in relation to economic conditions, in a village population of southern West Bengal, India. In: De Garine I, Harrison GA (1988) Coping With Uncertainty in Food Supply, Clarendon Press, Oxford
7. Bhattacharya N, Charaborty P, Chattopadhyay M, Rudra A (1991) How do the poor survive? Eco Pol Wkly 16:373–379
8. Khongsdier R (1995) A study on nutrition and health status in relation to some biosocial factors among the war Khasi of Meghalaya. PhD thesis in anthropology, North-Eastern Hill University, Shillong
9. Mondal B, Chattopadhyay M, Gupta R (2005) Economic condition and nutritional status: a micro level study among tribal population in rural West Bengal, India. Mal J Nutr 11(2):99–109
10. Banerji D (1982–1986) Poverty, class and health culture in India, vol 1 Prachi Prakashan, New Delhi
11. Basu BS, Buddhabdeb (1986) Tribal health socio-cultural dimensions. Inter-India Publications, New Delhi
12. Kar RK (1986) A note on health and sanitation among the tea labor in Assam. Vanyajati 36(2):1–9
13. Roy Burman BK (1986) Morbidity and nutritional status of scheduled tribes of India. In: Chaudhuri B (ed) Tribal health socio-cultural dimensions. Inter-India Publications, New Delhi
14. Sahu SK (1986) Social dimension of health of Tribals In India. In: Chaudhuri B (ed) Tribal health socio-cultural dimensions. Inter-India Publications, New Delhi

15. Basu S (1994) The state of the art—Tribal health. In: Basu S (ed) Tribal Health in India. Manak Publications, Delhi
16. Basu S (1996) Health and socio-cultural correlates in Tribal communities. In: Mann RS (ed) Tribes of India: ongoing challenges. M.D. Publications Pvt. Ltd., New Delhi
17. Basu S (1996) Need for action research for health development among the Tribal communities of India. S Asian Anthropol 17(2):73–80
18. Basu S (1996) Health status of Tribal women in India. In: Amar Kumar S, Jabli MK (eds) Status of Tribals in India—health, education and employment. Council for Social Development, Har Anand Publications, New Delhi
19. Basu S (1999) Dimensions of Tribal health in India, health and population. A Lecture Delivered at National Institute of Health and Family Welfare, New Delhi
20. Mahapatra S (1990) Cultuaral pattern in health care: a view from Santal world. In: Chaudhuri B (ed) Cultural and environmental dimension on health. Inter-India Publications, New Delhi
21. Ghosh PK, Uma CK, Kumar N (2004) Study of impact of Geogens as a deciding factor to control the spread of pathogens in and around Shankargarh, development block of Allahabad. M.Phil Thesis Awarded to P.K. Ghosh from A.P.S. University Rewa, Madhya Pradesh

Establishing the Concept of Biogeogens

Our Studies on Rural/Tribal Populations

The poor health status of any population is manifested not only in terms of morbidity and mortality rates but also in the capacity of individuals of a community to develop their human potentialities and meaningful productive life. Recognition of heath as an essential component of socio-economic development has led to the review and reorientation of health statistical services. Though healthcare facilities are overwhelmingly concentrated in rural areas of India, the 'socio-economic distance' prevents access for the rural poor. These socio-economic barriers include cost of health care, social factors, such as the lack of culturally appropriate services, language/ethnic barriers, and prejudices on the part of providers. Thus, there is a general belief that the rural/tribal populations in India are deprived of many facilities/healthcare services; and their nutritional status is also very low, in turn making them vulnerable to several diseases.

Now, the problem is how to prove this? While supervising the projects/thesis work [1–6] of the researchers in the Institute of Applied Sciences, Allahabad, several facts were revealed during our studies on health and nutritional status of the rural/tribal population of Shankargarh area, near Allahabad; which became the backbone of the concept of "Biogeogens."

Study area: Shankargarh block is situated in the Vindhyan region of India. The Vindhyan range is a low mountain range of Central India, extending in east–west direction from Varanasi to Gujarat State for a distance of 1,100 km. (700 miles). The range separates the drainage basin of the river Ganga in the north from the Deccan Plateau in the south. Elevations range from 450 to 500 m (1,500–3,000 ft.) and reach a maximum of 1,113 m (3,651 ft.). It includes South of Allahabad, Vindhyanchal, Mirzapur, Manigaon, Chakghat, Shahdol, Rewa, Panna, Satna, Umaria, Sidhi and adjoining areas. In 1948 Baghel Khand and Bundel Khand merged into Vindhya Pradesh, which, with several former enclaves of southern Uttar Pradesh, merged with Madhya Pradesh in 1956. The Vindhyan range is also known for its great potential for medicinal herbs.

N. Kumar, *Biogeogens and Human Health*, SpringerBriefs in Public Health,
DOI: 10.1007/978-81-322-1084-9_3,

In fact, our attention was earlier focused on studies on medicinal plants only; and as per earlier survey conducted by our researchers several plants were identified in the Vindhyan region, such as *Andrgraphis paniculata* (Fam. Acanthaceae), *Catharanthus roseus* (Apocynaceae), *Matthiola incana* (Fam. Brassicaceae), *Lagerstroemia speciosa* (Fam. Lyhraceae), *Althaea rosea* and *Sida cordifolia* (Fam. Malvaceae), *Butea monosperma* and *Tephrosia purpurea* (Fam. Papilionaceae), *Receda odorato* (Fam. Resedaceae), *Ruta graveolens* (Fam. Rutaceae), etc. However, gradually our attention shifted toward nutritional and health studies, as we came across several amazing facts, which finally forced to correlate our findings with other hidden realities unearthed during the extensive studies on the target population, leading to establishment of the Biogeogen concept.

Map of the study area

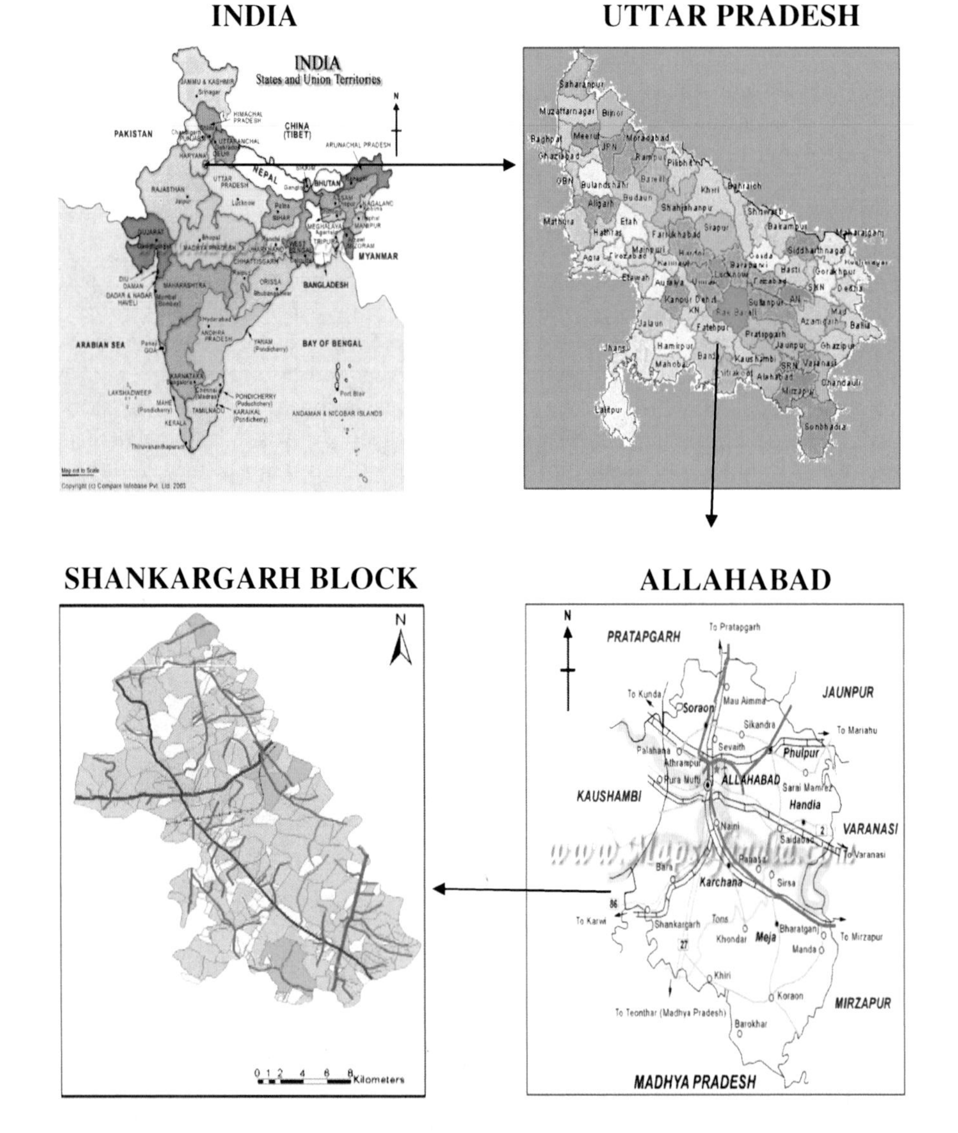

Work plan: As stated above, the story began with our ethnological studies, especially on anti-diabetic plants in the Vindhyan region. The preliminary survey of nearby areas of Shankargarh and the connected literature revealed that the people in and around this region were using several herbs/plants for the cure of diabetes and related ailments [4, 6, 7], such as Sadaphuli, Jarul, Gurmar, Sweet leaf, Neem, Babool, Bel, etc. It was also found that their knowledge of ethnomedicines although very rich, is without any experimental evidence. Therefore, a holistic study was planned under a project (supervised by the author and conducted by the researcher Mr. Prasanna Kumar Ghosh) in the Institute (IASc), to evaluate their (target population) knowledge of ethnomedicines and the overall status of their health to draw concrete inferences. Later, some of the important findings [3, 6–9] of this study were further extended and became a part of Mr. Ghosh's thesis, for the award of Ph.D. degree from the Forest Research Institute, Dehradun [10].

The systematic consultation of the literature on the aforesaid plants and diabetes was taken as the first step. It is a well-known fact as per prediction of World Health Organization (WHO) that developing countries will bear the brunt of diabetes in the twenty-first century, with 80 % of all new cases of this disease expected to appear in the developing countries by 2025. In one generation, diabetes has had a six-fold increase. In 1985 there were an estimated 30 million people with diabetes. Today, diabetes affects more than 230 million people, almost 6 % of the world's adult population. The number of people living with diabetes is expected to grow to 350 million in less than 20 years if action is not taken. Diabetes is one of the major causes of premature death worldwide. Every 10 s a person dies from diabetes-related causes. The death rates are predicted to rise by 25 % over the next decade. Diabetes is increasing faster in the world's developing economies than in developed countries. Seven out of ten countries with the highest number of people living with diabetes are in the developing world. With an estimated 35 million people with diabetes, India has the world's largest diabetes population. Each year another 6 million people develop diabetes. In many countries in Asia, in the Middle East, Oceania, and the Caribbean, diabetes affects 12–20 % of the adult population. Diabetes is a silent epidemic that claims as many lives each year as HIV/AIDS. In 2007, diabetes caused 3.5 million deaths globally. If the present trend persists, by 2025 the majority of people with diabetes in the developing countries will be in the 45–64 age group. Type 2 diabetes is responsible for 90–95 % of diabetes cases. 80 % of type 2 diabetes is preventable by changing diet, increasing physical activity, and improving the living environment (*WHO 1985*).

In developing countries, less than half of the people with diabetes are diagnosed. Without timely diagnosis and adequate treatment, complications and morbidity from diabetes rise exponentially. WHO estimates that diabetes, heart disease, and stroke together will cost about $555.7 billion loss in national income in China over the next 10 years, $303.2 billion in the Russian Federation; $336.6 billion in India; $49.2 billion in Brazil, and $2.5 billion even in a very poor country like Tanzania. These estimates are based on lost productivity, resulting primarily from premature death (International Diabetes Federation (*IDF), Diabetes Atlas, November 2006*).

The International Diabetes Federation, Diabetes Atlas released early in December in South Africa also confirms that there are around 40.9 million diabetics in India followed by China having around 39.8 million. It is estimated that in 2025, India will top the list, with 69.9 million diabetics, but China will put some distance between itself and India, having only 59.3 million diabetics. The increasing number of diabetics in India is a major challenge to the Indian economy. The numbers of diabetic patients are rapidly increasing all over the world, though the trends are different for different countries. In some places the growth rate is faster than others. Differences in lifestyle and the role of racial factors are the obvious reasons. In developed countries great importance is attached to physical activity and so diabetes has been brought under control to an extent. At the same time, in India, Pakistan, and Gulf countries, the number of diabetics is rapidly increasing. In 1995 every seventh diabetic person in the world was an Indian and by 2025 every fifth diabetic person will be an Indian. In 1995 the number of diabetics in India was 1.94 crores and by 2025 this number will swell to 5.70 crores. The number of diabetics is rapidly increasing in India, but what is more worrying is the fact that the younger age group is being more affected. At present 30 % of the diabetics are in the age group of 20–40 years.

The National Urban Diabetes Survey in India showed that more than 50 % of diabetes cases had their onset below the age of 50. Indians show significantly higher age-related prevalence when compared to the white population in the USA. Indians have several-fold higher prevalence of diabetes at all age groups in comparison with the European population as shown by the International Diabetes Epidemiology Group. It is also shown that risk of diabetes starts to increase at very low levels of BMI.

The risk of diabetes increases with small weight changes at a Body Mass Index above 22 kg/m^2. The cut-off value for healthy BMI in Indians is below 23 kg/m^2. It is interesting to note that the value for normal waist girth is also low in Indians (men 85 cm, women 80 cm). Despite having a lean BMI, an adult Indian has more chances of having abdominal obesity. The National Indian Survey showed that upper body adiposity was more common (50.3 %) than overweight as indicated by BMI (30.8 %). In Indians, central obesity shows a stronger association with glucose intolerance than generalized obesity. Studies from the UK and the USA have suggested that insulin resistance in non-obese Asians and Indians is due to the high percentage of visceral fat. This could also partly explain the higher prevalence of diabetes in them.

The higher insulin resistance may be partly due to a high body fat content seen in the Indian population. The body fat percentage of an Indian is significantly higher than his western counterpart with a similar BMI and blood glucose levels. According to a hypothesis, excess body fat and low muscle mass may explain the high prevalence of excess insulin in the blood and the high risk of type 2 diabetes in Indians. The major blood vessels leading to the heart and brain are affected as well as smaller ones leading to the kidneys and eyes are also affected because of chronic tissue complications.

Internal or external migration to a more affluent environment results in metabolic changes, resulting in higher blood glucose levels and related abnormalities.

The features of insulin resistance which include upper body adiposity and high body fat percentage as well as an abnormally high level of lipids and fats circulating in the bloodstream occur at a young age in Indians. Minor changes in weight or physical activity worsen insulin resistance. The new generation of children and adolescent show unprecedented levels of obesity. Several studies from India have highlighted that the epidemic of diabetes in urban India will become worse due to the rising trend of obesity in children. In urban south India, 16 % of school children are overweight and this shows a strong association with lack of physical activity and a high social stratum. Prevalence of type 2 diabetes in children is also increasing, probably due to less physical activity and altered dietary habit. Many studies on Indians have highlighted that the risk of cardiovascular diseases is high among Indians and could be related to the high prevalence of the metabolic syndrome. An epidemiological study showed that the prevalence of metabolic syndrome among urban adults was 41 %. It is interesting to note that although insulin resistance is common among Indians, it is not the major factor underlying the clustering for cardiovascular risk factors, namely high levels of blood glucose, higher lipid levels, hypertension, and upper body fat [11].

The high prevalence of diabetes has remained an urban phenomenon. So far all the previous epidemiological studies have illustrated a four-fold difference in the prevalence of diabetes between the urban and rural populations. But in present studies, the impact of socio-economic transition occurring in rural India, has shown a three-fold rise in the prevalence of diabetes in rural south India. This transition occurred during a period of 14 years and the prevalence has risen from 2.4 to 6.4 %. The contributing factors were the improved socio-economic status which encompassed an increasing family income and educational status, motorized transport, and a shift in the occupational structure. A similar situation is seen in the data available from neighboring countries such as Thailand, Malaysia, Bangladesh, and Pakistan. The Diabetes India Group has estimated the prevalence of diabetes in urban and rural areas in India. Urban areas were designated as those having a population of 100,000, and small towns and villages with a deemed population under 100,000 were designated as rural areas. The study reported an overall diabetes prevalence of 4.3 %: 5.6 % in urban areas and 2.7 % in rural areas. The total prevalence of Impaired Glucose Tolerance (IGT) was 5.2 %: urban being 6.3 % and rural 3.7 %. A 4.3 % of overall prevalence rate for diabetes in the Diabetes India Study confirms the WHO estimate of about 35 million adults with diabetes in India today. Now, with such an alarming rise in diabetic cases in India; and multifactorial pathogenesis of diabetes as discussed above, it seems a multimodel therapeutic approach is necessary to combat this disease.

Different medicinal systems are using the active plant constituents, discovered as natural hypoglycemic medicine, by virtue of traditional knowledge. Herbal drugs are considered free from side effects than synthetic ones. They are less toxic, relatively cheap, and popular. In India, medicinal plants have been used as natural medicine since the days of Vedic glory. Many of these medicinal plants and herbs are part of our diet as spice, vegetables, and fruits. Historically, in the "Atharv-Ved" (about 200 B.C.) a description of medicinal plants is made under a separate chapter "Ayurved." Sushruta (about 400 B.C.) compiled a classification of 700 herbal drugs under 37 classes in the "Sushruta Sanhita" (a compendium of ancient Indian surgery). Charak (about 600 B.C.) made the scientific classification of herbal drug plants based on remedial properties in his renowned treatise "Charak Sanhita" (a compendium of general medicine). The medicinal values of plants have been tested by trial-and-error method for a long time by different workers. Even today, great opportunities are still open for scientific investigations of herbal medicines for cure of diabetes and its complications.

Recently, plants and herbs are also being used as decoctions or in other extracted forms for their blood sugar lowering potential. There are some useful reviews on Indian medicinal plants having blood sugar lowering potential [12–14]. Many useful herbs introduced in pharmacological and clinical trials have confirmed their blood sugar lowering effect and repair of beta cells of *islets of Langerhans*. Details of some potent Indian herbs, their recently reported pharmacological and clinical hypoglycemic efficacy, active chemical constituents, their mechanism of action, and available toxicity status have been described by many.

Inspired by all these studies, and worried about the alarming situation of diabetic India, a vast survey was conducted in the region to gather information about the ethnological database as well as the health status of the target population.

The participatory observations revealed many interesting facts which were found in contrast to the views expressed by Basu and many others [15–20]. As per the firsthand information collected for this study it was found that the Kols and other tribes living in/around Allahabad (Shankargarh area) never live in isolation or remoteness. They regularly mix with other upper/lower castes in and around their areas; and a *Tribe—Caste—Continuum* (as discussed by Malinowaski [21]) has already been established giving rise to a changed outlook in these people (even in kol tribes), for the perception of developmental processes going on in the country.

The people of the target area

The terrain

The hutment

Although they depend mainly on agriculture and stone cutting work for earning their livelihood, no consistency in weekly occupational days were found; and overall it could be inferred on the basis of continuous participatory observation and interrogation that they get work for 4–5 days in a week earning approximately 110 Rs./day for their family (size around 6–7 persons). It could also be established that on average they work only for about 3–4 h. a day; see Table 1 (Appendix A).

Thus, on the basis of their work schedule (3–4 h. only per day; and 4–5 days in a week), their type of labor was categorized close to the range of moderate work.

The determination of the type of work/labor being performed by the said population was essential for estimating their nutritional requirements.

Nutritional studies: The extensive survey, data collection, and analysis of their diet and nutritional status were done; and for accuracy of results, the population was divided into male and female groups and further subdivided into different age groups, with the first group between 20 and 40 years. and the second group between 41 and 60 years (for both male and female).

The results were first tabulated on the basis of individual observations and then compiled for group analysis. Tables 2–5 (Appendix A) give the detailed account of their diet and nutritional status.

Anthropometric Analysis: The Body Mass Index of the population under study was derived using standard methods. The height and weight of total individuals from both sexes were calculated. The BMI was used as the cut-off point for assessment of CED (Chronic Energy Deficiency).

The calculations are given in Tables 6–9 (Appendix A).

Hematological Studies: After completing the nutritional and anthropometric studies it was also felt necessary to test the various hematological parameters of the population under study.

Blood samples were collected (male-57 and female-43) with the help of trained technician/medicos; and the results tabulated in Tables 10 and 11 (Appendix A).

The standard methods (Appendix B) were adopted for all the above investigations—nutritional, anthropometric, and hematological status analysis of the people of Shankargarh area.

Discussing the results in the context of Biogeogens: The nutritional values of food constituents and the corresponding energy status of each individual of both sexes of different age groups (as found and shown in Tables 2–5 (Appendix A)) clearly demonstrate that the nutritional requirements are almost within the standard range (as per ICMR recommendations). But, as the nutritional requirements differ from individual to individual, age to age, and sex to sex, it was felt necessary to compare the individual intake value of energy (calculated on the basis of nutritional data obtained for every individual) with the required energy value as per their individual age, sex, body weight and height, and type of work.

As against the reference weight (60 kg.) given in the table derived from the recommendations of *ICMR Advisory Committee 1989*, the average weight of men was calculated around 50 kg; similarly average weight for women was found around 45 kg against the weight of reference woman (50 kg). Therefore, it was not possible to calculate the Required Energy Value for the individuals, following the aforesaid ICMR recommendations. Hence, we searched for other options.

A person's metabolism varies with his physical condition and activity. A decrease in food intake can lower the metabolic rate as the body tries to conserve energy. Researcher Gary Foster, Ph.D., estimated that a very low calorie diet of fewer than 800 calories a day would reduce the metabolic rate by more than 10 % [22]. The oxidative system (aerobic) is the primary source of ATP supplied to the body at rest and during low intensity activities and uses primarily carbohydrates and fats as substrates. Protein is not normally metabolized significantly, except during long-term starvation and long bouts of exercise (greater than 90 min.) At rest, approximately 70 % of the ATP produced is derived from fats and 30 % from carbohydrates. Following the onset of activity, as the intensity of the exercise increases, there is a shift in substrate preference from fats to carbohydrates. During high intensity aerobic exercise, almost 100 % of the energy is derived from carbohydrates, if an adequate supply is available. Therefore, there is a direct correlation between the metabolic rate of an individual and his/her dietary intake/ requirements and type of activity performed. The primary organ responsible for regulating metabolism is the hypothalamus. The hypothalamus is located on the diencephalon and forms the floor and part of the lateral walls of the third ventricle of the cerebrum. The chief functions of the hypothalamus are:

1. Control and integration of activities of the autonomic nervous system (ANS).
   - The ANS regulates contraction of smooth muscle and cardiac muscle, along with secretions of many endocrine organs such as the thyroid gland (associated with many metabolic disorders).
   - Through the ANS, the hypothalamus is the main regulator of visceral activities, such as heart rate, movement of food through the gastrointestinal tract, and contraction of the urinary bladder.
2. Production and regulation of feelings of rage and aggression.
3. Regulation of body temperature.
4. Regulation of food intake, through two centers:
   - The feeding center or hunger center is responsible for the sensations that cause us to seek food. When sufficient food or substrates have been received and leptin is high, then the satiety center is stimulated and sends impulses that inhibit the feeding center. When insufficient food is present in the stomach and ghrelin levels are high, receptors in the hypothalamus initiate the sense of hunger.
   - The thirst center operates similarly when certain cells in the hypothalamus are stimulated by the rising osmotic pressure of the extracellular fluid. If thirst is satisfied, osmotic pressure decreases.

All of these functions taken together form a survival mechanism that causes us to sustain the body processes that Basal Metabolic Rate (BMR) and Resting Metabolic Rate (RMR) measure. Thus estimation of BMR/RMR may provide clues for the required energy for sustaining the life processes; and both basal metabolic rate and resting metabolic rate are usually expressed in terms of daily rates of energy expenditure. The early work of the scientists J. Arthur Harris and Francis G. Benedict showed that approximate values could be derived using body surface area(computed from height and weight), age, and sex, along with the oxygen and carbon dioxide measures taken from calorimetry. Therefore, the Basal Metabolic Rate for each individual was calculated as per their age, height, and body weight following the Harish-Benedict Equation [23] (and its refinement in terms of Resting Metabolic Rate (RMR) as described by Mifflin Equation [24]); and further multiplied with the activity factor [23] as per their type of activity level:

The original equations from Harris and Benedicts are:

$$\text{For men, } P = \left(\frac{13.7516\,\mathrm{m}}{1\,\mathrm{kg}} + \frac{5.0033\,h}{1\,\mathrm{cm}} - \frac{6.7550\,a + 66.4730}{1\,\mathrm{year}}\right)\frac{\mathrm{kcal}}{\mathrm{day}}$$

$$\text{For women, } P = \left(\frac{9.5634\,\mathrm{m}}{1\,\mathrm{kg}} + \frac{1.8496\,h}{1\,\mathrm{cm}} - \frac{4.6756\,a + 655.0955}{1\,\mathrm{year}}\right)\frac{\mathrm{kcal}}{\mathrm{day}}$$

where P is total heat production at complete result, m is the weight, h is the stature (height), and a is the age. The difference in BMR for men and women is mainly due to the difference in their body weights.

It was the best prediction equation until recently, when MD Mifflin and ST St Jeor in 1990 created a new equation:

$$P = \left(\frac{9.99\,\mathrm{m}}{1\,\mathrm{kg}} + \frac{6.25\,h}{1\,\mathrm{cm}} - \frac{4.92\,a + s}{1\,\mathrm{year}}\right)\frac{\mathrm{kcal}}{\mathrm{day}},$$

where s is +5 for males and -161 for female.

To calculate the daily calorie needs, this BMR/RMR value is multiplied by a factor with a value between 1.2 and 1.9, depending on the person's activity level, as given below:

1. for sedentary work (little or no exercise) : Calorie Calculation = BMR $\times$ 1.2
2. for light activity (light exercise/sports 1–3 days/week) : Calorie Calculation = BMR $\times$ 1.375
3. for moderate activity (moderate exercise/sports 3–5 days/week) : Calorie Calculation = BMR $\times$ 1.55
4. for very active (hard exercise/sports 6–7 days/week) : Calorie Calculation = BMR $\times$ 1.725
5. for extra active (very hard exercise/sports and physical job or training) : Calorie Calculation = BMR $\times$ 1.9

Thus, the required energy values for every individual were calculated, keeping them in the category of moderate activity workers (as per calculation in Table 1 (Appendix A); and compared with the intake value of energy. Amazingly, it was found that they are not at all energy deficient. Their required energy values were found almost equal to the intake values of energy. Therefore, it was inferred that both men and women of the said population have adequate nutrition and energy for sustaining their life processes and labor. Now again, the question which must not be left unanswered is, was it possible to calculate the exact energy requirement without considering the geo-socio-psycho factors? Actually, it was the participatory approach, which allowed keen observation of their daily routine, cultural ethos, food habits, behavioral syndromes, etc., leading to the determination of the types of food they eat as well as the type of activities they perform; and without these minute calculations it was never possible to reach such conclusion in terms of arithmetic perfection of energy required and energy spent. This is how the Biogeogens play their role.

However, the nutritional and energy data alone are not sufficient to determine the exact status of their health. Therefore, data were cross-examined with their physical and physiological health status. Their physical status was determined with the help of anthropometric analysis (Tables 6–9 (Appendix A)); and their Body Mass Index (BMI) was found well within the normal range, i.e., 18.5–24.9, which once again confirmed that they had nutrition within the standard range; and no Chronic Energy Deficiency (CED) or any sign of overnutrition was existing.

Further, to give no chance to speculations the physiological status was also examined after taking the consent of all 100 individuals for hematological studies. Blood samples of 100 individuals (including both sexes of different age groups) were analyzed for the estimation of Blood group, Hemoglobin, Random Glucose, S. Urea, S. Creatinine, S. Calcium, S. Cholesterol, S. Triglyceride, S. Sodium, S. Potassium, etc. The results obtained (Tables 10 and 11 (Appendix A)) show that the blood parameters of all the individuals (except a few, who were diagnosed as diabetic) were within the normal range showing normal physiological status, which is not possible without the balance in the supply of nutrition and corresponding energy expenditure.

Now after cross-examining the energy/nutritional, physical, and physiological status of health, the study of disease pattern was also tabulated. They were found mainly affected by communicable diseases due to lack of sanitation facilities, low awareness level, and also absence of preventive and promotive healthcare services (an essential component of Biogeogens). Their family history indicated that only a few died due to cancer in their immediate ancestral generation (to cross-check, the PHC data were also obtained). The analysis of results was discussed in the International Symposium on "Diet and stress relation in causation and prevention of cancer" held at ITRC, Lucknow [25]; in which It was reported that they are mainly dependent on a vegetarian diet and also they are not subject to undernutrition or overnutrition. The disease pattern also depicted that their immune function is stronger; and these findings are in consonance with others [26–29]. It has also been well established that several phytonutrients are the best protectors against cancer [30, 31] and the effect of calorie restriction for protection against cancer has also been proved in several studies [32–35]; our deeper analysis of the nutritional data further established that because they do not get more food than the required values, their calorie restriction is automatically imposed upon keeping their BMI well within the normal/healthy range. Those who were getting slightly higher food intake regularly and not spending their energy accordingly, could not maintain healthy status and detected diabetic with higher BMI and Glycemic index. It was also found that their ethnological knowledge and practices play a vital role in prevention and cure of several diseases, as a few who were found diabetic, their random blood sugar count was not very high even though they were not taking any medicine. But they were found using herbs for the cure of their diabetic condition; and that too in a functional correlation with the environmental factors [36]—again a scientific proof of the role of Biogeogens in determining the health status, earlier described by us in detail. Similar use of such herbs has been reported by several others [37–39]; and it has been reported that the juice of fresh leaves of such plants indicated a significant antidiabetic activity in albino rabbits and the reduction in blood glucose was also found dependent on the periodicity of admixture doses. Thus, a complete and comprehensive approach is a must while studying the status of health of an individual/community; especially the consideration of Biogeogens could solve several intricacies/problems of health care.

Another important finding was the ratio of the blood groups, which significantly revealed that the population has a direct link with the central Indian tribes as well as the lower castes of the region. As per similar findings in another study [40], the percentage of the AB blood group was found lower, while B blood group was higher, depicting the possibilities of racial mixing.

However, no correlation could be established with the postulates of Dr. D'Adamo, who believes our blood group determines how our bodies deal with different nutrients. His theory [41] is based on the idea that each blood group has its own unique antigen marker (a substance that the body recognizes as being alien) and this marker reacts badly with certain foods, leading to all sorts of potential health problems. Furthermore, Dr. D'Adamo believes that levels of stomach acidity and digestive enzymes are linked with our blood type. Consequently, he says, by following a diet designed specifically for your blood type, your body digests and absorbs food more efficiently, with the result that you lose weight (http://www.drlam.com/blood_type_diet/ [42]). Dr. D'Adamo believes that because blood types evolved at different times throughout history, we should eat a diet based on the types of foods our ancestors typically ate at the time when our blood type was first recognized!

As per his theory, the blood group O was the first blood type to be identified, although how we know this is anyone's guess—we are talking about our hunter-gatherer ancestors who were around in 50,000 B.C! Dr. D'Adamo believes because our type O ancestors survived and thrived on a high-protein, meat-based diet, that is the type of diet blood group Os should follow. Next came the emergence of blood type A, some time around 15,000 B.C! By this time, our ancestors' hunter-gathering days were over and instead they started to settle into farming-type communities. The creation of blood type A around this time meant our ancestors did well on a vegetarian-based diet. Again, Dr. D'Adamo recommends that blood group As should today follow a veggie diet. Further, blood type B supposedly evolved around 10,000 B.C—thanks to our nomadic ancestors. They left their farms and started wandering the land, constantly moving from place to place. Consequently, Dr. D'Adamo's theory is, blood group Bs today can get away with eating a varied diet that consists of most foods including meat, dairy, grains, and vegetables. Finally, came blood type AB, which evolved just 1,000 years ago! Dr. D'Adamo thinks this blood type helped our ancestors make the transition to modern times. Meaning that people with blood group AB can eat a mixture of the foods suitable for both blood group A and blood group B.

In our studies, the population was found surviving on almost the same type of diet without any specific complaints/diseases, hence Dr. D'Adamo's claim could not be justified in case of our studies; however, a clear-cut correlation was established in their nutritional requirements and energy expenditure. The balance of energy metabolism was the major factor in maintaining good health; while their ethnological knowledge also plays an important role in combating diseases, such as—an interesting and important use of *Catharanthus roseus* was observed by our group. The said population used sadaphuli (*Catharanthus roseus*) with jarul (*Lagerstroemia speciosa*) for curing diabetes. Admixture-dependent doses (ADD) are prepared from these two plants in combination with two other plants—anjan (*Ailanthus excelsa*) and nux vomica (*Stychnos nux vomica*). Further, a specific time/period was also maintained in taking formulation/doses of these plants. All these are worth investigating further in relation to genetical/immunological/biochemical studies on the said population to explore many untouched realities and their scientific details.

Over all and above, our studies have evidently established that three dimensions, e.g., environmental, population, and cultural, together play a significant role in deciding the status of human health.

Therefore, the following measures are suggested, which could be helpful in adopting preventive and promotive healthcare strategies, especially for those above 40 years of age, having normal health and residing in subtropical-humid region of the globe:

- One should adopt *Nature-Man-Medico* complex, to lead a healthy life, i.e., Learn to live with the environment by judicious and eco-friendly utilization of the natural resources around one's surrounding.
- Always take a balanced diet; and as much food as required for meeting the daily energy expenses only. Do not overeat (calculation of energy required could be done as per formula described above in the discussion section of the text after determining the activity factor; and energy value of the food be calculated as per standard table of ICMR or table published in the book of Dr. M. Swaminathan-Essentials of Food and Nutrition, The Bangalore Printing and Publishing Co.Ltd., Bangalore).
- Take seasonal and fresh fruits and vegetables (after washing carefully; do not overcook) regularly in your diet.
- Eat as much fibrous food as possible.
- Consume at least 2–3 L of potable water.
- Do not take extra salt or sugar in your diet.
- Fix-up and maintain a daily routine of food intake; avoid fasting as well as frequent or intermittent eating.
- Say no to tobacco and alcohol; and take natural (ethno) medicines as far as possible for curing minor ailments.
- Take 3–4 cups of green tea without sugar and milk; but with a small piece of ginger and one clove bud. May add fresh lemon juice for flavor and instant freshness.

- One red chilly, one red tomato, and one red carrot be taken daily as a salad with a small onion and 3–4 buds of garlic, mixed with five drops of coriander oil and a little salt (this acts as an anti-oxidant dose) during the break fast.
- Do regular exercises, at least for half an hour on all weekdays; If possible walk in-between your sedentary working hours and use bicycle for commuting.
- Avoid air-conditioned environment as much as possible.
- Never overexert or take over anxiety.
- Meditate at least for 5–10 min daily in the morning or night.
- Avoid excessive use of cosmetics and synthetic clothes.
- Be cheerful and playful even while doing serious things.
- Have faith in yourself and be positive.

Supporting Evidences from the Evolutionary History of Man

The evolution, i.e.,"Descent with Modification," is an ongoing process. Some variation takes place in nature, adaptation follows leading to modification in the genotype and phenotype, selection pressure exerts resulting in the survival of the fittest and if there is no further dilution for some period a new species comes into existence. But this all is not simple!

Biogeogens play a decisive role in determining the path of evolutionary mechanism. So is the case for Human Evolution. But before going into the details, a brief description of human evolution may be required and helpful for readers of this book to understand the role of different components of Biogeogens, which played a pivotal role in deciding the fate of our evolutionary history.

The problem of human evolution lies in:

- Its correlation and coexistence with many other primates, especially the apes,
- as well as its dissimilarity with them; and also with its own varieties.

Actually the story begins as early as 35 million years ago, in the Oligocene epoch of the tertiary period of the Cenozoic era.

- It was the period of great apes—like Parapithecus/Propliopithecus which gave rise to early apes like Proconsuls in Africa at/around 20 million years ago.
- Further, these apes moved to Eurasia giving rise to Dryopithecus in Europe and Sivapithecus in Asia.
- Dryopithecus moved back to Africa giving rise to African Apes and hominid varieties like Sahalenthropus –7 million years ago.

- Thus, a direct link detected till date is *Orrorin tugenensis* and *Sahelanthropus tchadensis* (dated at 7–6 million years old).
- Hence, evolution of man could be directly linked and compared with the apes on one hand and with *Sahelanthropus on the other.*

Further, how the different varieties of Hominids evolved and the divergence took place is better understood only after a close and comparative study of the fossil evidences. On the basis of the details of the fossil evidences as described in the recent literature [43–50], a phylogenetic tree has been prepared, specially pinpointing the major shift in the course of evolutionary divergence through the ages (Fig. 1)

For having a glimpse into the major evolutionary developments, a comparative chart has also been prepared tracing the evolutionary links between one and the other species, supposed to be close to the homo varieties.

Fossils	Cranium size (c.c.)	Behaviour
Sahelan thropus tchadensis	350	Very close to chimpanzees. Not truly biped Hominid or not—still debatable
Australopithecus africanus	420–500	Strong sexual dimorphism Bipedalism
Australopithecus robustus	530	May be the first to use tools (but could not make it)
Homo habilis	650	Oldwan tool culture (could make blunt stone tools, axes without hand)
Homo erectus (or Homo ergaster)	750	(*erectus* for Eurasian varieties; *ergaster* for African) Acheulean tool culture (Levalloisian technique for making core by percussion flaking); widespread use of fire
Homo heidelbergensis (also classified as *Homosapiens archaic*)	1,200	Became specialized in making cores and flakes. More use of flakes was started
Homo neanderthalensis	1,450	Mousterian tool culture (middle Palaeolithic culture; more flakes to prepare sharper tools) band, housing, religious belief
Homo sapiens	Slightly less than naenderthals	Aurignacian tool culture (upper Palaeolithic culture) sophisticated tools and language

Further, a pictographic view of some of the fossils is given (sketches drawn on the basis of skull diagrams available on the website: http://www.talkorigins.org/faqs/homs/) to make the differences more visible

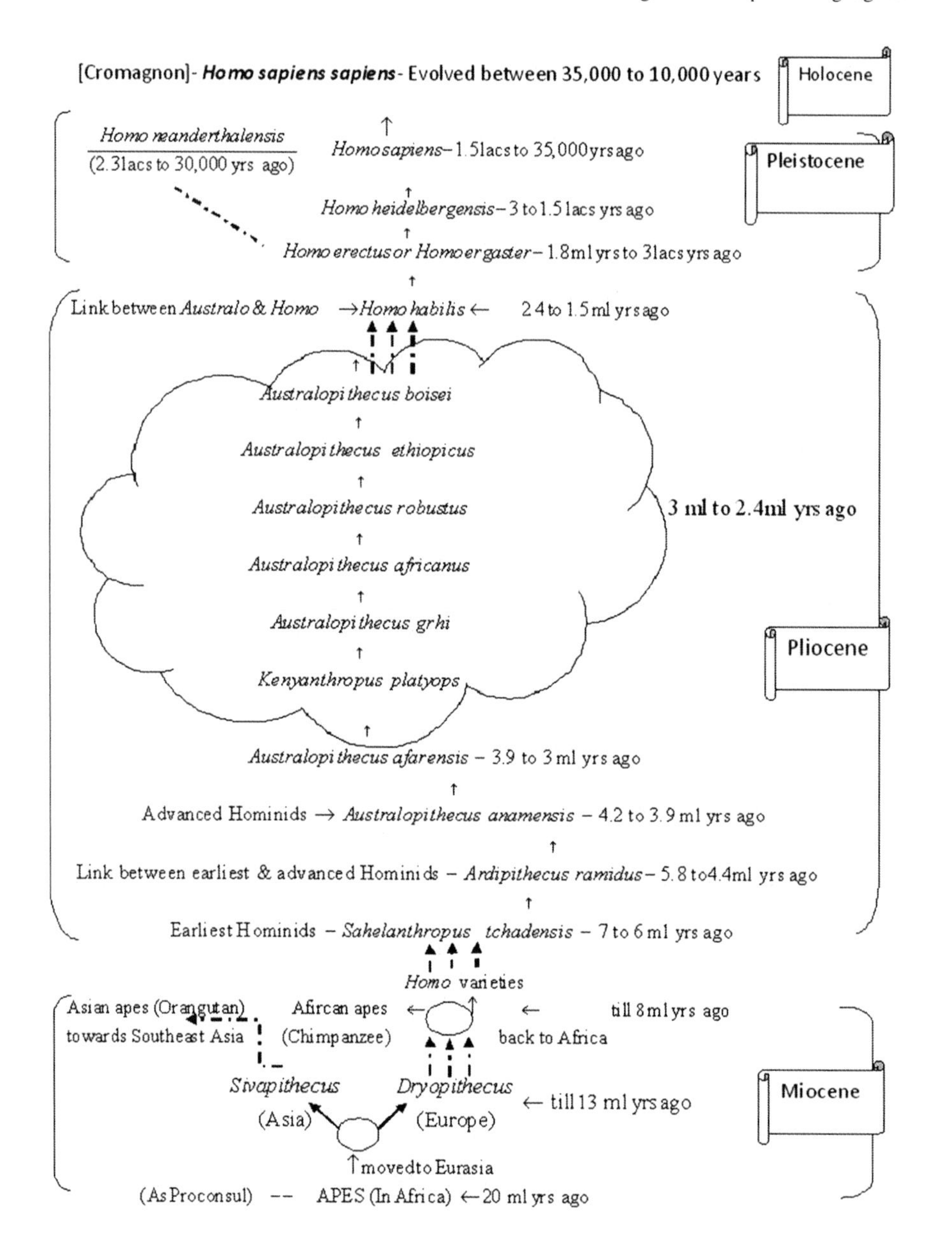

Fig. 1 Phylogenetic evolutionary tree of man. Abbreviations: *ml* million, *yrs* years

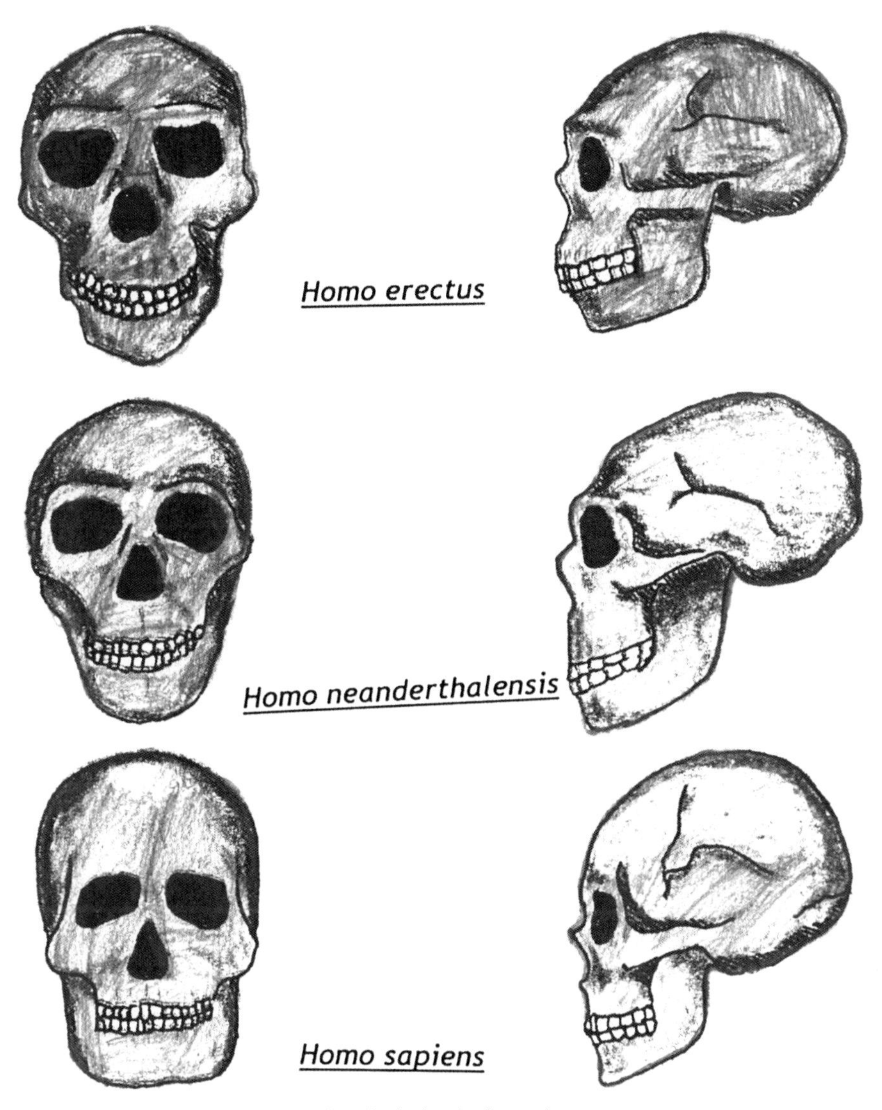

Now, from all the above details it is obvious that:

- Every time the brain grew incrementally, good things happened. Thus we can propose that advantage may be related to increased capacity (Cranial Capacity) for an aspect of intelligent behavior.
- From *Sahelanthropus* to *Homo sapien neanderthalensis* locomotion (perfection in bipedalism) was perhaps the cause for allowing space for the increase in cranium's size. Because, Natural selection adapts the form of the pelvis to the dominant form of locomotion which affects the size of the birth canal and hence fetal cranial capacity.

- Further, cephalopelvic disproportion in parturition was the major limiting factor in the evolution of cranial capacity. From *Neanderthal* to *Homo sapien*, small reduction in Cranium size, was due to the limiting role of the birth canal. But as the behaviour kept on evolving, the area needed to store and analyze information/experience was also to increase—which was compensated by the increase in the folds of the cerebral cortex, i.e., increase in the area of cerebral cortex.
- This increase in surface area (folding) was correlated with the cultural development (as speech/language and other behavioral adaptations)—as proved by the presence and absence of—Broca's area (Center for Speech) in the fossil evidences.
- Casts of brain impression (by liquid latex) have also shown the presence of Broca's area in *Homo habilis*—although it was not so developed, as in the case of modern man.

This explanation about our advancements clearly depicts how our geographical domain influenced the cultural, leading to anatomical alterations and further evolutionary changes; supporting the hypothesis of the role of Biogeogens in affecting our growth and development.

Further, there are other interesting facts deep in the history of the Ice ages. The present ice age, defined as the Pliocene-Quaternary glaciation, started about 2.58 million years ago during the late Pliocene; and remarkably this was also the time of the major split between the Austrlo and Homo varieties. Since then, the world has seen cycles of glaciation with ice sheets advancing and retreating on 40,000- and 100,000-year timescales called glacial periods, glacials or glacial advances, and interglacial periods, interglacials or glacial retreats. The earth is currently in an interglacial, and the last glacial period ended about 10,000 years ago; that was the time since we evolved completely as the modern man.

Didier Paillard [51] suggests that the late Pleistocene glacial cycles can be seen as jumps between three quasi-stable climate states. The jumps are induced by the orbital forcing, while in the early Pleistocene the 41,000-year glacial cycles resulted from jumps between only two climate states. A dynamical model explaining this behavior was proposed by Peter Ditlevsen [52]. During the course of evolution major changes also took place between these three quasi-stable climate states of late Pleistocene. Why such drastic changes occurred, could be explained as—within the ice ages, more temperate and more severe periods occur. The colder periods are called *glacial periods*, the warmer periods *interglacials*, such as the Eemian Stage. Glacials are characterized by cooler and drier climates over most of the Earth and large land and sea ice masses extending outward from the poles. Mountain glaciers in otherwise unglaciated areas extend to lower elevations due to a lower snow line. Sea levels drop due to the removal of large volumes of water above sea level in the icecaps. There is evidence that ocean circulation patterns are disrupted by glaciations. During the glacial and interglacial stages the redistribution of ice-water on the surface of the Earth and the flow of mantle rocks causes changes in the gravitational field as well as changes to the distribution of the moment of inertia of the Earth. These changes result in influencing the adaptation pattern of the animals and humans also.

Not only this, the human behavior has also influenced drastically the atmosphere. At a meeting of the American Geophysical Union (December 17, 2008), scientists detailed evidence in support of the controversial idea that the introduction of large-scale rice agriculture in Asia, coupled with extensive deforestation in Europe began to alter world climate by pumping significant amounts of greenhouse gases into the atmosphere over the last 1,000 years. In turn, a warmer atmosphere heated the oceans making them much less efficient storehouses of carbon dioxide and reinforcing global warming, possibly forestalling the onset of a new glacial age [53].

There is considerable evidence that over the very recent period of the last 100–1000 years, the sharp increases in human activity, especially the burning of fossil fuels, has caused the parallel sharp and accelerating increase in atmospheric greenhouse gases which trap the sun's heat. The consensus theory of the scientific community is that the resulting greenhouse effect is a principal cause of the increase in global warming which has occurred over the same period, and a chief contributor to the accelerated melting of the remaining glaciers and polar ice. A 2012 investigation finds that dinosaurs released methane through digestion in a similar amount to humanity's current methane release, which "could have been a key factor" to the very warm climate 150 million years ago [54].

Some scientists believe that the Himalayas are a major factor in the current ice age, because these mountains have increased Earth's total rainfall and therefore the rate at which CO_2 is washed out of the atmosphere, decreasing the greenhouse effect [55]. The Himalayas' formation started about 70 million years ago when the Indo-Australian Plate collided with the Eurasian Plate, and the Himalayas are still rising by about 5 mm/year because the Indo-Australian plate is still moving at 67 mm/year. The history of the Himalayas broadly fits the long-term decrease in the Earth's average temperature since the mid-Eocene, 40 million years ago. Again, that was the period when major divergence took place in the evolutionary pattern of the mammals.

Thus it could be inferred that there is a direct correlation between the environmental, cultural, and biological factors affecting each other in an integrated manner; and when there is any maladaptive interaction between these, the problem erupts causing even irreparable damage to the system.

Evolution of Alternative System of Medicines

How forceful the theory of Biogeogens is? Nothing could explain it better than the Evolution of Alternative System of Medicines!

As per Shirazi, the allopathic system of medicine's approach is mostly focused on the presented signs and symptoms, and it treats the manifestations of what might be a deeper and more chronic condition. A good example of this conclusion is when a patient visits most conventional clinics around the country with a headache. They will check her vital signs, ask some questions, and will send her home with some pain medication as a form of treatment. If she returns later with the same complaint, they will probably order some basic blood work, or they might even order a CAT scan of her head to rule out a brain tumor; ultimately they will send her home with more or stronger pain meds. If she returns again with the same complaint, they will then refer her to a brain specialist or a pain clinic, where she might receive more pain-related medicines or therapies. The main point of referring to this common condition as an example of standard care in most conventional (allopathic) clinics is to demonstrate the philosophy of standard care and the main focus of this type of approach. Thus, the allopathic system of medicine refers to the practice of conventional medicine that uses pharmacologically active agents or physical interventions (like surgery) to treat or suppress symptoms or patho-physiologic processes of disease [56].

Although allopathic medicine likes to define everything, including water as toxic or as a poison, there are in reality certain medicinal substances, including water, that have *no side effects* even at high doses. Thus, effective medicine (maintaining good health or returning to it) is caught up with the question of how to fix what ails us without suffering side effects (poisonous effects) from the medicines we use. There is no healing system more powerful than that which employs Nature's primordial substances, materials so pure and close to nature that they yield benefits without the typical side effects of pharmaceutical drugs. In the twenty-first century the center of pharmacology needs to be shifted away from medicines that add to people's already heavy toxic burdens, to medicines and protocols that reduce these burdens [57].

That was the cause giving rise to the need for an alternative system of medicine; although even today an alternative system of medicine is considered as any of a wide range of healthcare practices, products, and therapies, which typically are not included in the degree courses of established medical schools; examples include homeopathy, Ayurveda, chiropractic and acupuncture, etc. [58]. In respect of alternative medicine since the 1970s in North America, Great Britain, and elsewhere, there has been a tendency for the terms 'alternative' and 'complementary' to be used interchangeably to describe a wide diversity of therapies that attempt to use the self-healing powers of the body by amplifying natural recuperative processes to restore health (… and that is how the dogma of Biogeogens works, balancing the mechanism of coherent interaction of bio and geo gens for eliminating the pathogens). By 1990, approximately 60 million Americans had used one or more complementary or alternative therapies to address health issues, according to a nationwide survey in the United States published in 1993 by David Eisenberg [59]. A study published in the November 11, 1998 issue of the Journal of the American Medical Association reported that 42 % of Americans had used complementary and alternative therapies, up from 34 % in 1990 [60].

Alternative medicine practices and beliefs are diverse in their foundations and methodologies, and typically make use of preparations and dosages other than such as are included in the Pharmacopeia, recognized by established medical schools. Methods may incorporate or base themselves on traditional medicine, folk knowledge, spiritual beliefs, or newly conceived approaches claiming to heal [61, 62]. African, Caribbean, Pacific Island, Native American, and other regional cultures have traditional medical systems as diverse as their diversity of cultures [63]. In India also several types of alternative system of medicines/therapies are in practice. A short list is given below:

- Worship to God/Supernatural Power (Prayer) to heal is the most common alternative medicine in India and many western countries; and faith healing is a part of many religions.
- Some believe that meditation affects health. Transcendental meditation is the good example, which originated in India; and is being practiced in many parts of the globe.
- Yoga as a healing practice involves stretching, exercise, and meditation related to the Hindu religion, and makes claims to healing in the spiritual realm. Crores of Westerners are now following difficult Yogic practices.
- The Ayurvedic medicine of India is a complex system developed over thousands of years, based on Vedic scripture and regional supernatural belief systems and traditional use of herbs, etc.
- Acupuncture is a part of Traditional Chinese medicine in which needles are inserted in the body to alter the flow of plasma/energy believed to propel the blood and influence health.
- Chiropractic manipulation of the spine was developed in the United States, and involves manipulating the spine to cure diseases.
- Homeopathy was developed in Europe; and is now spread all over the world.
- Magnets and light are used in therapies based on the cumulative effect of electromagnetism on an individual's health.
- Dietary supplements are now in full practice; and several neutraceuticals are in the world market.

Thus, Alternative medical systems are complete health systems with their own approaches to diagnosis and treatment that differ from the conventional biomedical approach to health. Some are cultural systems such as Ayurveda and Traditional Chinese Medicine, while others, such as Homeopathy and Naturopathy are relatively recent and were developed in the West [64]. The evolution of Ayurveda is based on three prime factors—cough (*KAUGH*), gas (*VATA*), and bile (*PITTA*). The Ayurvedik remedial measures rely on the coherent and proportionate presence of these three factors in our body, and any disturbance in their ratio may give rise to organic problems or pathogens, just as the maladaptive interaction of 'biogens' or 'biogens and geogens' give rise to pathogens. Similarly, Naturopathy is based on a belief in vitalism, which posits that a special energy called vital energy or vital force guides bodily processes such as metabolism, reproduction, growth, and adaptation [65]. Naturopathy favors a holistic approach with noninvasive treatment and, similar to conventional medicine, encourages minimal use of surgery and drugs. The term "naturopathy" is derived from Greek and Latin, and literally translates as "nature disease" [66]. Modern naturopathy grew out of the Natural

Cure movement of Europe [67, 68]. The term was coined in 1895 by John Scheel and popularized by Benedict Lust [69], the "father of U.S. naturopathy" [70]. Beginning in the 1970s, there was a revival of interest in the United States and Canada in conjunction with the holistic health movement [71]. Today, naturopathy is primarily practiced in the United States and Canada [72]. All these, either Ayurveda or Naturopath, simply describes our faith in nature or interaction of biological system with the environmental one, interwoven on the matrix of the respective traditions (culture) in a systematic manner; and that is what has been explained in the earlier part of this book as Biogeogens in a holistic way.

Whether they stand on the validity scale of science or not, could simply be understood by the fact that about two-thirds of the world's population is treated by such alternative systems of medicine. Considering the importance of these practices, WHO has recognized its value and included these in their program to achieve the goal of "Health for all by 2000 AD." Not only that, the National Center for Complementary and Alternative Medicine, (NCCAM) USA, budget has been on a sharp sustained rise to support complementary medicines.

References

1. Ghosh PK, Chauhan KU, Kumar N (2004) Study of impact of geogens as a deciding factor to control the spread of pathogens in and around Shankargarh, Development Block of Allahabad,M. Phil. Thesis Awarded to P.K. Ghosh from A.P.S. University Rewa, M.P
2. Vartika, Niralaya S, Sinha A, Kumar N (2003) Study of medicinal plants being used by Kols of Allahabad region of Uttar Pradesh. Presented in 73rd Annual Session of NASI held at Ahamadabad
3. Vartika, Prasanna GK, Singh KP, Sinha A, Kumar N (2004) Vitamin deficiency and immunosuppresion in Kols. Presented in 74th Annual Session of NASI held at University of Rajasthan, Jaipur, 2–4 Dec 2004
4. Vartika, Singh KP, Sinha A, Kumar N (2004) Admixture dependent doses prepared from the medicinal plants by "Kol" tribe for treatment of diabetes. Natl Acad Sci Lett 27(7, 8):257–260
5. Vartika, Sinha A, Singh KP, Kumar N (2005) Role of trace elements in maintenance of nutritional balance. In: Proceeding of 75th annual session of NASI held at Central University, Pondicherry, 8–9 Dec 2005
6. Vartika, Ghosh KP, Singh KP, Sinha A, Kumar N (2006) Promotive health care strategies for kols of Shankargarh, Allahabad. Presented in 30th annual session of Indian social science congress held at Aligarh Muslim Universiy, Aligarh, 2006
7. Ghosh KP, Kumar A, Kumar N (2005) Diabetes cured by Sadaphuli. Presented in 75th annual session of NASI held at Central University, Pondicherry, 8–9 Dec 2005
8. Ghosh KP, Kumar A, Vartika PA, Singh KP, Kumar N (2009) Study of the geography of health in and around Shankargarh, Allahabad to adopt promotive health care strategies. Natl J Life Sci 6(2):173–176
9. Ghosh KP, Kumar A, Vartika PA, Singh KP, Kumar N (2009) Malnutrition and it's effects on Kols. Natl J Life Sci 6(3):346–350
10. Ghosh KP (2013) Ph.D.Thesis, awarded by the F. R. I., Dehradun
11. Ramchandran A, Mahajan V (2005) India has the largest and fastest growing diabetic population in the world. Spectr. Tribute

12. Mukherjee PK, Saha K, Pal M, Saha BP (1997) Effect of Nelumbo nucifera rhizome extract on blood sugar level in rats. J Ethnopharmacol 58(3):207–213
13. Grover JK, Yadav S, Vats V (2002) Medicinal plants of India with anti-diabetic potential. J Ethnopharmacol 80(1): 81–100
14. Saxena A, Vikram NK (2004) Role of selected Indian plants in management of type 2 diabetes. J Altern Complement Med 10(2):369–378
15. Basu S (1994) The state of the art – tribal health. In: Basu S (ed) Tribal health in India. Manak Publications, Delhi
16. Basu S (1996) Health and socio-cultural correlates in tribal communities. In: Mann RS (ed) Tribes of India: ongoing challenges. M.D. Publications Pvt. Ltd., New Delhi
17. Basu S (1996) Need for action research for health development among the tribal communities of India. S Asia Anthropol 17(2):73–80
18. Basu S (1996) Health status of tribal women in India. In: Singh AK, Jabli MK (eds) Status of tribals in India – health, education and employment. Council for Social Development, Har Anand Publications, New Delhi
19. Basu S (1999) Dimensions of tribal health in India, health and population. A Lecture Delivered at National Institute of Health and Family Welfare, New Delhi
20. Mahapatra S (1990) Cultuaral pattern in health care: a view from Santal World. In: Chaudhuri B (ed) Cultural and environmental dimension on health. Inter-India Publications, New Delhi
21. Malinowski B (1945) The dynamics of culture change. Yale University Press, New Heaven
22. Whitman S (2003) The truth about metabolism. Shape. Sept 2003
23. Harris JA, Benedict FG (1918) A biometric study of human basal metabolism. Proc Natl Acad Sci USA 4(12):370–373
24. Mifflin MD, St Jeor ST, Hill LA, Scott BJ, Daugherty SA, Koh YO (1990) A new predictive equation for resting energy expenditure in healthy individuals. Am J Clin Nutr 51:241–247
25. Kumar N (2005) Diet and stress relation in causation and prevention of cancer. In: Lecture delivered in international symposium held at ITRC, Lucknow in March 2005
26. Frentzel–Beyme R, Chang-Claude J (1994) Vegetarian diet and colon cancer : the German experience. Am J Clin Nutr 59(suppl) 1143S–1152S
27. Kinlen LJ, Hermon C, Smith PG (1983) A proportionate study of cancer mortality among members of a vegetarian society. Br J Cancer 48:355–361
28. Malter M, Schriever G, Eilber U (1989) Natural killer cells, vitamins, and other blood components of vegetarian and omnivorous men. Nutr Cancer 32:271–278
29. Block G, Patterson B, Subar A (1992) Fruit, vegetables, and cancer prevention: a review of the epidemiological evidence. Nutr Cancer 18:1–29
30. Tannenbaum A (1942) The genesis and growth of tumors, II. Effects of caloric restriction per se. Cancer Res 2:460–467
31. Macrae FA (1993) Fat and calories in colon and breast cancer: from animal studies to controlled clinical trials. Prev Med 22:750–766 (review)
32. Risch HA, Jain M, Marrett LD, Howe GR (1994) Dietary fat intake and risk of epithelial ovarian cancer. J Natl Cancer Inst 86:1409–1415
33. Cramer DW, Welch WR, Hutchison GE et al (1984) Dietary animal fat in relation to ovarian cancer risk. Obstet Gynecol 63:833–835
34. Hill HA, Austin H (1996) Nutrition and endometrial cancer. Cancer Causes Control 7:19–32 (review)
35. Moerman CJ, De Mesquita HBB, Runia S (1993) Dietary sugar intake in the aetiology of biliary tract cancer. Int J Epidemiol 22:207–214
36. Vartika, Singh KP, Sinha A, Kumar N (2004) Admixture dependent doses prepared from the medicinal plants by "Kol" tribe for treatment of diabetes. Natl Acad Sci Lett 27(7, 8):257–260
37. Nammi S, Boini KM, Lodagala D, Srinivas Behara S, Babu R (2003) The juice of fresh leaves of *Catharanthus roseus Linn*. Reduces blood glucose in normal and alloxan diabetic rabbits. BMC Complement Altern Med 3:4

38. Singh SN, Vats P, Suri S, Shyam R, Kumria MML, Ranganathan S, Sridharan K (2001) Effect of an antidiabetic extract of *Catharanthus Roseus* on enzymic activities in Streptozotocin induced diabetic rats. J Ethnopharmacol 76:269–277
39. Chattopadhyay RR, Sarkar SK, Ganguli S, Banerjee RN, Basu TK (1991) Hypoglycemic and antihyperglycemic effect of leaves of *Vinca rosea Linn*. Ind J Physiol Pharmacol 35:145–151
40. Macfarlane EWE, Sarkar SS (1941) Blood groups in India. http://deepblue.lib.umich.edu/bitstream
41. D'Adamo P (with additional material by Catherine Whitney) (2002). The eat right 4 Your Type complete blood type encyclopedia. Riverhead. ISBN 1-57322-920-2
42. http://www.drlam.com/blood_type_diet/
43. Kumar N, Sinha A (2002) on 'Evolution of Man'. In: Proceedings of the 72nd annual session of the national academy of sciences, India, Allahabad
44. Kumar N et al. (2005) on 'Increased cranial ......expertise'. In: Proceedings of the national symposium on recent advances in neurobiology, held at A. U., Allahabad
45. Begun DR (2003) Planet of the Apes'. Sci Am 289(2):64–73
46. Jared D (1997) Guns, germs and steel: a short history of everybody for the last 13,000 years pub. Vintage
47. Galik K, Senut B, Pickford M, Gommery D, Treil J, Kuperavage AJ, Eckhardt RB (2004) External and internal morphology of the BAR 1002'00 Orrorin tugenensis femur. Science 305(5689):1450–1453
48. Jeff H (2004) Donkey domestication began in Africa New Scientist. newscientist.com. Accessed 17 June 2004
49. Richmond BG, Strait DS (2002) Evidence that humans evolved from a knuckle-walking ancestor. Nature 404:382–385
50. White T, Asfaw B, Degusta D, Gilbert H, Richards G, Suwas G, Clark Howell F (2003) Pleistocene homo sapiens from Middle Awash, Ethiopia. Nature 423:742–747
51. Paillard D (1998) The timing of Pleistocene glaciations from a simple multiple-state climate model. Nature 391(6665):378–381. Bibcode 1998 Natur.391. 378P. doi:10.1038/34891
52. Ditlevsen PD (2009) Bifurcation structure and noise-assisted transitions in the Pleistocene glacial cycles. Paleoceanography 24:PA3204. Bibcode 2009PalOc. 24.3204D. doi:10.1029/2008PA001673
53. Did early climate impact divert a new glacial age? (2008) Newswise. Newswise.com/articles/view/547541
54. Davies E (2012). BBC nature – Dinosaur gases 'warmed the Earth'. Bbc.co.uk. http://www.bbc.co.uk/nature/17953792
55. Raymo ME, Ruddiman WF, Froelich PN (1988). Influence of late Cenozoic mountain building on ocean geochemical cycles. Geology 16(7):649–653. Bibcode 1988 Geo. 16. 649R. doi:10.1130/0091-7613(1988)016<0649:IOLCMB>2.3.CO;2
56. Shirazi SF (2012) Allopathic medicine vs. holistic medicine. Acupuncture Today 13(09)
57. http://www.naturalallopathic.com/cms/
58. http://en.wikipedia.org/wiki/Alternative_medicine
59. Eisenberg DM et al (1993) Unconventional medicine in the United States – prevalence, costs, and patterns of use. N Engl J Med 328(4):246–252
60. Eisenberg DM, Davis RB, Ettner SL et al (1998) Trends in alternative medicine use in the United States, 1990–1997: results of a follow-up national survey. JAMA 280(18):1569–1575
61. Sampson W (1995) Antiscience trends in the rise of the 'Alternative Medicine' Movement. Ann N Y Acad Sci 775:188–197
62. Beyerstein BL (2001) Alternative medicine and common errors of reasoning. Acad Med 76(3):230–237
63. What is complementary and alternative medicine (CAM)? National Center for Complementary and Alternative Medicine. http://nccam.nih.gov/health/whatiscam/. Retrieved 11 July 2006
64. Goldrosen MH, Straus SE (2004) Complementary and alternative medicine: assessing the evidence for immunological benefits. Nat Perspect 4:912–921

65. Sarris J, Wardle J (2010) Clinical naturopathy: an evidence-based guide to practice. Elsevier, Chatswood
66. Naturopathy: an introduction (PDF). National Center for Complementary and Alternative Medicine, 2007. http://nccam.nih.gov/health/naturopathy/D372.pdf
67. Brown PS (1988) Nineteenth-century American health reformers and the early nature cure movement in Britain. Med Hist 32(2):174–194
68. Langley S (2007) History of naturopathy. Excerpt from the naturopathy workbook. UK: College of Natural Medicine (CNM). http://www.naturopathy-uk.com/blog/2007/11/28/history-of-naturopathy/
69. Report 12 of the council on scientific affairs (A-97). American Medical Association, 1997. http://www.ama-assn.org/ama/no-index/about-ama/13638.shtml
70. Baer HA (2001) The sociopolitical status of U.S. naturopathy at the dawn of the 21st century. Med Anthropol Q 15(3):329–346
71. Frey R (April 1) Naturopathic medicine. Encyclopedia of medicine. Gale (Cengage). http://findarticles.com/p/articles/mi_g2601/is_0009/ai_2601000954
72. A study of costs and length of stay of inpatient naturopathy-evidence from Germany. Complement Ther Clin Pract, 2011. http://www.ncbi.nlm.nih.gov/pubmed/21457898

Appendix A

Tables A.1, A.2, A.3, A.4, A.5, A.6, A.7, A.8, A.9, A.10 and A.11

Table A.1 Calculation of weakly working hours

	Male	Female
Average working hrs. as per questionnaire	8 h	8 h
Resting period during the working hrs.	1 h (Lunch)	1.43 h (Lunch)
Time consumed in drinking water, tobacco chewing and smoking	0.75 h	0.50 h
Toilet and other activities	0.50 h	0.50 h
Total resting hrs.	2.25 h	2.43 h
Real working hrs. (Average working hrs. − Total resting hrs.)	8−2.25 = 5.75 h	8−2.43 = 5.57 h
Total working days in a weak (4–5 days)	4.5 days	4.5 days
Average working hrs. per day	Total working days in a weak × Real working hrs. / 7	
	4.5 × 5.75 / 7 = 3.70 h	4.5 × 5.57 / 7 = 3.58 h

Excerpts of the results obtained during the studies (the work was further extended by Dr. Prasanna K. Ghosh, in the I.A.Sc., Allahabad for his Ph.D Thesis under the guidance of the author)

N. Kumar, *Biogeogens and Human Health*, SpringerBriefs in Public Health,
DOI: 10.1007/978-81-322-1084-9, © The Author(s) 2013

Table A.2 Nutritional analysis for males (No. 30, age: 20–40)

S. No.	Age (Yr.)	Wt. (Kg)	Prot. (gm)	Fat (gm)	Carb. (gm)	Ca (mg)	Fe (mg)	Mg (mg)	Na (mg)	K (mg)
1.	26	50	64.96	18.60	416.68	285.03	24.25	725.91	2432.07	1209.70
2.	27	47	67.20	19.05	436.68	266.04	24.11	762.23	2509.54	1225.37
3.	32	49	64.25	18.95	421.71	263.44	23.69	734.93	2547.18	1152.8
4.	32	49	64.13	17.70	406.95	268.50	23.62	724.32	2592.19	1301.97
5.	29	55	70.15	19.80	461.96	277.46	25.30	813.65	2628.15	1231.85
6.	33	42	71.18	21.99	393.45	284.54	23.39	679.51	2665.50	1167.50
7.	30	37	56.61	15.72	355.91	255.42	21.55	630.29	2701.26	1106.71
8.	35	45	73.19	19.23	467.54	287.10	26.31	819.58	2756.08	1404.76
9.	28	55	70.90	19.15	453.31	291.07	25.87	790.92	2834.25	1382.87
10.	29	63	69.11	18.92	448.20	273.84	24.89	784.04	2866.49	1270.60
11.	33	47	71.28	23.05	393.49	282.08	23.30	677.57	2900.69	1163.77
12.	35	44	63.24	18.82	410.17	255.34	22.96	709.50	2507.57	1169.44
13.	33	57	66.87	19.44	431.06	254.94	25.70	758.75	2988.76	1296.84
14.	35	53	68.98	19.69	454.85	270.38	24.91	800.97	3060.48	1218.12
15.	33	65	67.72	19.85	440.16	271.96	24.54	770.16	3140.43	1245.64
16.	28	52	73.91	22.29	406.93	292.00	24.29	703.87	3219.24	1224.60
17.	26	47	66.98	18.78	428.70	271.13	24.21	744.95	2434.15	1268.61
18.	27	56	68.08	19.96	437.18	273.51	24.72	762.84	2513.78	1273.83
19.	28	64	67.99	17.94	437.59	281.12	25.10	765.31	2592.35	1258.39
20.	35	50	65.08	18.57	419.74	262.87	23.55	730.5	2665.95	1193.72
21.	35	50	73.07	23.94	407.34	281.31	23.72	704.2	2745.98	1223.79
22.	30	54	64.95	17.57	415.44	272.81	23.92	722.05	2825.75	1215.71
23.	28	54	70.75	22.52	399.16	270.75	22.84	689.48	2897.34	1122.01
24.	35	35	68.49	21.36	379.43	264.42	21.89	650.71	2974.24	1098.18
25.	36	55	69.02	20.03	447.85	284.84	25.48	785.21	3066.59	1300.47
26.	27	45	63.24	17.46	412.11	272.21	23.35	713.68	3138.00	1197.87

(continued)

Table A.2 (continued)

S. No.	Age (Yr.)	Wt. (Kg)	Prot. (gm)	Fat (gm)	Carb. (gm)	Ca (mg)	Fe (mg)	Mg (mg)	Na (mg)	K (mg)
27.	29	43	62.11	17.51	408.54	253.05	22.52	709.57	3211.87	1117.71
28.	29	49	73.54	19.24	460.93	352.41	29.52	812.54	2449.37	1421.18
29.	31	52	64.82	18.26	415.49	271.35	23.86	720.17	2511.04	1219.42
30.	29	49	60.64	17.9.0	390.98	252.76	21.81	670.45	2584.17	1160.59

S. No	Carotene (μg)	Thiamine (mg)	Riboflavin (mg)	Niacin (mg)	Total B6 (mg)	Folic Acid (μg)	Vitamin C (mg)	Choline (mg)	Energy (Kcal)
1.	2320.94	2.01	1.26	20.65	2.07	132.96	36.81	158.05	2093.22
2.	1778.33	2.11	1.27	21.76	2.15	129.93	31.11	154.83	2186.26
3.	2081.84	2.02	1.25	21.00	2.08	128.00	33.44	140.77	2113.67
4.	1800.88	2.04	1.14	20.83	2.05	134.89	36.53	183.74	2042.65
5.	1963.05	2.19	1.31	23.03	2.29	131.55	33.31	156.58	2305.86
6.	2250.72	1.95	1.32	19.53	1.94	152.80	34.64	150.26	2055.76
7.	1998.06	1.78	1.06	18.04	1.81	122.74	32.36	139.18	1790.88
8.	1661.64	2.35	1.35	23.75	2.33	149.61	30.98	158.15	2335.09
9.	1880.06	2.28	1.33	22.97	2.25	149.02	34.61	154.27	2268.32
10.	1874.72	2.18	1.3	22.46	2.22	136.55	33.34	156.85	2238.75
11.	2247.94	1.94	1.32	19.50	1.94	151.21	33.04	146.73	2065.85
12.	1855.37	1.99	1.22	20.41	2.01	126.81	31.05	142.72	2062.24
13.	1892.04	2.10	1.24	21.49	2.14	143.05	17.97	176.39	2165.89
14.	1873.53	2.17	1.30	22.76	2.25	132.42	33.55	154.57	2271.76
15.	1873.86	2.13	1.28	22.02	2.17	134.19	33.52	158.44	2209.40
16.	2259.63	2.04	1.36	20.36	2.02	159.90	33.65	146.69	2123.25
17.	1873.17	2.11	1.27	21.45	2.12	136.47	33.15	157.58	2150.98
18.	1873.45	2.16	1.29	21.98	2.17	138.89	32.17	150.73	2199.93
19.	2096.68	2.15	1.30	21.97	2.18	138.66	33.00	148.57	2183.01
20.	1867.36	2.03	1.25	20.9	2.07	129.94	32.06	151.20	2105.67

(continued)

Table A.2 (continued)

S. No	Carotene (μg)	Thiamine (mg)	Riboflavin (mg)	Niacin (mg)	Total B6 (mg)	Folic Acid (μg)	Vitamin C (mg)	Choline (mg)	Energy (Kcal)
21.	1931.99	2.04	1.33	20.34	2.00	154.29	31.28	150.94	2136.41
22.	2094.66	2.04	1.26	20.75	2.06	135.64	34.95	149.01	2079.00
23.	2104.65	1.93	1.31	19.68	1.96	143.8	32.70	151.26	2081.64
24.	2013.77	1.85	1.27	18.63	1.85	141.73	32.05	148.73	1983.30
25.	2106.81	2.19	1.32	22.57	2.22	142.98	37.87	164.84	2246.87
26.	2085.99	1.99	1.23	20.43	2.02	128.75	37.08	156.19	2057.77
27.	1856.74	1.94	1.21	20.20	1.99	121.96	33.78	146.82	2039.48
28.	3321.22	2.33	1.41	23.41	2.39	162.4o	37.10	144.60	2310.04
29.	2089.52	2.03	1.25	20.72	2.06	133.51	33.64	149.37	2084.79
30.	1859.57	1.88	1.18	19.22	1.90	122.46	35.45	163.00	1966.86

See Table A.1 footer

Table A.3 Nutritional analysis for males (No. 27, age: 41–60)

S.No.	Age (Yrs.)	Wt. (Kg)	Prot. (gm)	Fat (gm)	Carb. (gm)	Ca (mg)	Fe (mg)	Mg (mg)	Na (mg)
31.	**55**	**60**	**71.24**	**31.43**	**475.00**	**263.67**	**22.04**	**646.17**	**2744.12**
32.	42	55	70.81	22.07	400.83	282.70	23.34	691.34	2430.91
33.	45	40	60.27	18.49	341.89	258.57	20.64	604.44	2971.12
34.	50	45	64.50	21.35	355.98	275.79	21.53	604.48	2975.78
35.	**50**	**62**	**74.62**	**32.67**	**495.00**	**261.81**	**22.85**	**692.08**	**3059.37**
36.	50	42.	63.33	16.68	404.25	244.43	22.98	724.47	3142.88
37.	50	30	50.71	17.09	331.85	217.34	18.28	556.04	2414.87
38.	47	61	69.03	21.92	400.31	277.35	23.08	691.99	2623.32
39.	**50**	**60**	**73.14**	**30.48**	**488.21**	**287.29**	**24.22**	**738.54**	**2822.39**
40.	50	30	49.38	16.18	317.19	217.90	18.34	553.63	2414.59
41.	50	47	68.30	23.29	369.15	281.47	22.17	628.78	2938.02
42.	50	45	60.79	19.93	384.99	236.90	19.84	566.50	2929.72
43.	55	47	61.84	17.25	397.87	257.19	22.26	685.11	2427.70
44.	55	45	58.78	17.17	374.56	253.20	21.23	640.44	2700.81
45.	55	48	66.93	20.22	363.44	260.50	22.03	648.59	2552.60
46.	52	52	64.68	18.25	414.96	269.80	23.44	719.32	2627.77
47.	51	48	65.77	22.22	356.04	269.26	21.11	605.24	2697.02
48.	55	45	59.62	21.84	324.29	249.46	19.34	546.86	2768.96
49.	51	47	66.43	22.88	355.60	265.20	20.89	602.22	2894.72
50.	52	50	65.52	21.28	359.60	270.52	21.80	639.82	2817.04
51.	55	55	62.19	18.94	397.91	254.03	22.45	686.01	3057.51
52.	54	48	61.98	22.88	335.76	254.14	19.75	566.73	3006.54
53.	51	46	63.95	21.27	356.75	247.88	20.34	606.15	3087.41
54.	55	39	54.74	20.95	330.24	380.35	19.06	552.90	2824.17

(continued)

Table A.3 (continued)

S.No.	Age (Yrs.)	Wt. (Kg)	Prot. (gm)	Fat (gm)	Carb. (gm)	Ca (mg)	Fe (mg)	Mg (mg)	Na (mg)
55.	**55**	**60**	**72.23**	**29.78**	**487.37**	**277.67**	**24.25**	**744.25**	**2670.18**
56.	**60**	**61**	**73.75**	**30.43**	**495.78**	**266.57**	**23.24**	**716.69**	**2508.66**
57.	60	45	54.81	16.42	355.49	228.08	19.77	603.08	2457.68

S.No.	K mg)	Carotene (µg)	Thiamine (mg)	Riboflavin (mg)	Niacin (mg)	Total B6 (mg)	Folic acid (µg)	Vitamin C (mg)	Choline (mg)	Energy (Kcal)
31.	**1168.63**	**2079.72**	**1.87**	**1.18**	**18.72**	**1.84**	**128.85**	**36.93**	**146.30**	**2467.83**
32.	1222.99	2001.58	1.99	1.30	19.88	1.96	149.30	34.40	162.47	2084.53
33.	1027.97	2115.19	1.67	1.10	17.12	1.71	129.76	36.25	151.61	1774.46
34.	1096.82	2308.81	1.76	1.24	17.43	1.73	142.14	36.04	142.97	1873.36
35.	**1205.41**	**1788.66**	**1.95**	**1.12**	**19.84**	**1.97**	**130.36**	**32.39**	**152.6**	**2572.51**
36.	1176.78	1597.63	1.98	1.12	20.61	1.89	121.74	33.10	153.68	2019.81
37.	939.720	1778.76	1.55	1.05	15.92	1.57	101.70	32.16	131.44	1683.43
38.	1135.91	2206.90	1.93	1.29	19.72	1.96	140.97	34.57	150.47	2073.94
39.	**1175.11**	**2217.28**	**2.04**	**1.34**	**20.99**	**2.08**	**146.17**	**35.91**	**153.89**	**2519.72**
40.	936.650	1656.30	1.53	0.95	15.74	1.56	101.02	28.85	126.84	1611.30
41.	1156.61	2246.50	1.82	1.27	18.05	1.80	146.25	34.80	161.90	1958.82
42.	1018.76	1872.49	1.63	1.16	16.31	1.63	125.61	28.84	151.20	1962.21
43.	1171.39	1862.72	1.92	1.20	19.61	1.94	124.32	34.64	160.89	1993.38
44.	1133.28	1860.03	1.81	1.16	18.35	1.81	121.91	35.72	157.73	1887.22
45.	1107.45	1945.69	1.81	1.17	18.44	1.83	140.30	39.17	158.47	1902.94
46.	1226.16	1868.73	2.02	1.24	20.62	2.04	131.09	33.93	157.15	2082.04
47.	1058.55	2244.75	1.72	1.24	17.26	1.72	138.70	34.64	152.30	1886.69
48.	906.290	2359.73	1.53	1.18	15.54	1.55	127.43	34.97	135.11	1731.64
49.	1093.77	2029.11	1.75	1.24	17.31	1.71	141.72	34.10	155.83	1893.42
50.	1072.11	2167.78	1.78	1.17	18.17	1.81	142.15	36.51	153.35	1891.34

(continued)

Table A.3 (continued)

S.No.	K mg)	Carotene (μg)	Thiamine (mg)	Riboflavin (mg)	Niacin (mg)	Total B6 (mg)	Folic acid (μg)	Vitamin C (mg)	Choline (mg)	Energy (Kcal)
51.	1171.14	1863.87	1.94	1.21	19.72	1.95	127.34	33.24	154.74	2010.21
52.	966.570	2235.55	1.59	1.19	16.11	1.61	130.18	34.84	146.71	1796.39
53.	999.180	1999.52	1.72	1.22	17.41	1.71	133.70	32.90	133.76	1873.62
54.	1055.54	2174.20	1.58	1.14	15.85	1.64	110.68	36.07	150.78	1756.83
55.	**1287.87**	**1877.91**	**2.10**	**1.27**	**21.36**	**2.11**	**136.47**	**34.38**	**165.80**	**2506.42**
56.	**1217.11**	**1871.28**	**2.00**	**1.23**	**20.47**	**2.03**	**129.04**	**33.81**	**163.43**	**2551.85**
57.	1001.88	1835.14	1.67	1.10	17.21	1.71	107.36	30.01	139.70	1788.33

See Table A.1 footer

Table A.4 Nutritional analysis for females (No. 15, age: 20–40)

S.No.	Age (Yr.)	weight (Kg)	Protein (gm)	Fat (gm)	Carb. (gm)	Ca (mg)	Fe (mg)	Mg (mg)	Na (mg)
1.	40	44	51.05	18.70	295.11	326.98	17.73	496.51	3152.27
2.	40	49	54.96	15.97	345.27	246.60	20.74	608.08	3210.76
3.	36	40	49.38	17.32	306.73	232.52	18.28	508.36	2613.62
4.	40	45	53.48	17.82	331.03	234.45	19.08	554.67	2694.45
5.	39	40	47.10	17.44	303.32	208.84	17.10	503.43	2765.12
6.	36	40	48.40	17.72	306.23	214.51	17.23	505.35	2806.23
7.	40	43	50.94	15.72	316.57	239.12	19.49	552.11	3090.08
8.	40	40	49.57	15.83	301.63	215.69	17.23	492.63	2415.89
9.	39	43	49.63	16.89	311.37	217.86	17.85	516.06	2453.85
10.	40	48	58.55	21.54	319.34	241.36	18.61	530.42	2494.86
11.	40	50	54.25	15.09	332.06	240.93	20.12	579.68	2542.69
12.	40	40	46.82	16.53	294.79	203.68	16.36	480.97	2568.47
13.	40	46	47.19	17.64	297.36	218.87	17.21	488.70	2689.02
14.	37	46	58.62	18.04	317.97	254.03	19.83	554.90	2774.23
15.	39	35	43.25	15.42	274.01	202.39	15.49	443.61	2958.65

S.No.	K (mg)	Carotene (μg)	Thiamine (mg)	Riboflavin (mg)	Niacin (mg)	Total B6 (mg)	Folic Acid (μg)	Vitamin C (mg)	Choline (mg)	Energy (Kcal)
1.	1160.00	1689.42	1.53	1.09	14.51	1.51	115.14	31.48	147.25	1552.09
2.	1095.28	1860.52	1.73	1.03	17.46	1.74	119.44	32.22	143.71	1743.92
3.	966.330	2058.04	1.48	1.04	14.60	1.47	107.24	32.91	136.19	1579.82
4.	1041.47	1850.88	1.59	1.08	15.84	1.59	110.52	32.44	159.16	1697.90
5.	866.690	1821.35	1.42	1.01	14.37	1.43	97.470	31.54	126.18	1558.23
6.	937.400	1739.62	1.44	1.01	14.43	1.43	98.320	31.65	147.29	1577.47
7.	1041.37	1990.58	1.58	0.98	15.85	1.59	114.17	33.81	148.71	1610.89
8.	1011.49	1614.79	1.45	1.00	14.21	1.42	102.65	29.78	160.90	1546.76

(continued)

Table A.4 (continued)

S.No.	K (mg)	Carotene (μg)	Thiamine (mg)	Riboflavin (mg)	Niacin (mg)	Total B6 (mg)	Folic Acid (μg)	Vitamin C (mg)	Choline (mg)	Energy (Kcal)
9.	947.090	1829.79	1.47	1.03	14.76	1.48	101.00	29.21	141.18	1595.45
10	995.420	1970.00	1.53	1.13	15.15	1.51	122.36	32.97	157.37	1704.92
11.	1118.70	1729.27	1.69	1.01	16.73	1.67	119.13	30.38	152.67	1680.34
12.	917.450	1599.42	1.38	0.98	13.80	1.37	94.610	30.30	145.82	1514.73
13.	927.820	1917.17	1.40	1.00	14.01	1.40	100.01	33.34	142.97	1536.49
14.	1040.13	2118.75	1.58	1.06	15.79	1.59	128.64	33.81	161.25	1668.16
15.	849.570	1822.85	1.25	0.95	12.60	1.25	89.980	34.70	151.61	1407.38

See Table A.1 footer

Table A.5 Nutritional analysis for females (No. 28, age: 41–60)

S. No.	Age (Yr.)	Wt (Kg)	Prot. (gm)	Fat (gm)	Carb. (gm)	Ca (mg)	Fe (mg)	Mg (mg)	Na (mg)
16.	42	38	48.49	23.02	289.65	413.53	16.70	464.57	2778.52
17.	42	40	50.47	16.67	309.39	299.67	18.91	535.84	2932.63
18.	42	49	52.28	17.30	329.50	228.03	19.02	549.93	2419.40
19.	42	40	50.89	15.74	306.23	240.79	19.09	531.96	2499.27
20.	43	42	47.64	15.76	305.22	215.16	17.36	504.03	2964.68
21.	45	45	49.37	17.98	314.70	231.09	18.47	525.54	3203.55
22.	45	43	54.29	19.14	298.94	222.56	17.37	493.52	2883.21
23.	43	40	47.01	17.77	271.76	209.30	15.95	439.89	3037.48
24.	43	38	47.40	18.66	263.32	154.58	12.83	418.79	3031.03
25.	45	50	58.77	18.17	366.49	264.10	21.55	623.95	2428.33
26.	45	44	58.62	21.10	338.56	238.10	19.33	573.46	2452.62
27.	45	40	49.27	17.70	306.79	230.22	18.08	510.44	2573.24
28.	42	36	46.68	15.85	297.77	218.99	17.41	490.34	2650.43
29.	**50**	**60**	**71.71**	**29.83**	**487.58**	**240.47**	**17.59**	**471.60**	**2728.71**
30.	50	45	54.50	18.65	287.47	215.93	17.39	495.39	2805.89
31.	**52**	**58**	**70.43**	**32.27**	**503.92**	**230.89**	**18.47**	**523.93**	**2930.90**
32.	50	50	56.84	16.25	355.68	262.58	21.72	630.14	3054.52
33.	50	45	54.32	19.24	282.50	241.18	17.97	485.14	3202.07
34.	**55**	**57**	**72.83**	**30.23**	**495.74**	**233.79**	**18.07**	**486.16**	**2418.45**
35.	50	52	55.71	18.19	345.82	250.74	20.07	576.18	2468.05
36.	50	45	55.62	21.41	297.48	227.43	17.25	488.50	2569.18
37.	50	35	46.09	17.67	290.11	216.82	17.12	475.27	2691.18
38.	**60**	**63**	**73.35**	**31.43**	**483.14**	**209.67**	**16.34**	**461.15**	**2765.73**
39.	50	40	44.15	16.26	277.27	214.90	16.27	449.94	2922.83
40.	50	50	59.50	22.24	322.25	243.30	18.72	538.53	3083.71

(continued)

Table A.5 (continued)

S. No.	Age (Yr.)	Wt (Kg)	Prot. (gm)	Fat (gm)	Carb. (gm)	Ca (mg)	Fe (mg)	Mg (mg)	Na (mg)
41.	50	35	42.24	13.78	260.49	207.06	15.98	443.88	3196.71
42.	**50**	**60**	**75.02**	**29.57**	**490.83**	**211.92**	**15.09**	**400.08**	**2406.08**
43.	50	39	40.66	17.00	255.13	206.06	15.21	406.54	2802.25

	K (mg)	Carotene (μg)	Thiamine (mg)	Riboflavin (mg)	Niacin (mg)	Total B6 (mg)	Folic acid (μg)	Vitamin C (mg)	Choline (mg)	Energy (Kcal)
16.	980.180	2040.34	1.41	1.03	13.82	1.44	98.71	32.21	116.48	1559.20
17.	1007.80	2051.77	1.53	0.97	15.37	1.57	107.01	31.71	134.01	1588.87
18.	1012.29	1829.72	1.61	1.07	15.94	1.58	110.24	29.15	128.44	1682.26
19.	1043.24	1999.92	1.53	0.97	15.16	1.54	113.42	33.72	158.06	1569.60
20.	926.810	1823.83	1.45	1.02	14.53	1.43	101.41	33.24	134.65	1552.72
21.	950.280	2055.48	1.51	1.05	15.12	1.50	108.84	35.31	128.60	1617.58
22.	884.310	1949.48	1.41	1.09	14.11	1.41	114.83	31.47	137.40	1584.76
23.	820.73	2022.90	1.27	1.00	12.61	1.26	101.68	32.73	132.91	1434.57
24.	811.750	578.110	1.25	0.94	12.19	1.13	96.13	27.05	149.31	1410.50
25.	1174.33	2092.33	1.79	1.16	17.92	1.79	125.96	37.38	170.32	1863.89
26.	888.590	2175.52	1.57	1.16	16.23	1.62	117.84	31.82	124.64	1778.15
27.	954.270	2067.30	1.45	1.04	14.52	1.46	105.55	35.41	152.31	1583.06
28.	911.280	2047.16	1.41	1.01	14.12	1.40	101.58	34.36	134.54	1519.89
29.	**910.600**	**2188.04**	**1.39**	**1.09**	**13.57**	**1.35**	**121.66**	**35.67**	**135.49**	**2505.63**
30.	913.970	1667.07	1.43	0.99	14.19	1.42	118.60	27.50	141.61	1535.29
31.	**1057.09**	**1772.67**	**1.52**	**0.95**	**15.02**	**1.51**	**110.90**	**32.94**	**163.96**	**2587.35**
32.	1106.89	2134.90	1.76	1.06	17.92	1.81	121.55	33.13	144.40	1795.66
33.	955.860	2126.49	1.41	1.01	13.87	1.40	124.41	35.33	153.42	1519.95
34.	**974.620**	**2192.75**	**1.44**	**1.03**	**14.07**	**1.41**	**108.76**	**35.35**	**142.60**	**2546.35**

(continued)

Table A.5 (continued)

	K (mg)	Carotene (μg)	Thiamine (mg)	Riboflavin (mg)	Niacin (mg)	Total B6 (mg)	Folic acid (μg)	Vitamin C (mg)	Choline (mg)	Energy (Kcal)
35.	1219.07	1865.33	1.71	1.10	16.80	1.64	122.22	39.80	197.74	1768.98
36.	913.820	1979.07	1.40	1.09	13.93	1.39	117.19	33.95	156.57	1604.64
37.	935.500	1964.30	1.40	1.00	13.80	1.36	104.59	35.65	143.28	1503.24
38.	**902.910**	**1828.65**	**1.33**	**0.98**	**13.22**	**1.32**	**96.93**	**33.47**	**149.40**	**2508.83**
39.	887.300	2051.24	1.29	0.97	12.87	1.29	96.61	36.95	150.77	1431.48
40.	968.460	1991.71	1.54	1.15	15.35	1.52	125.02	33.96	151.98	1726.64
41.	858.680	1742.49	1.28	0.86	12.68	1.27	96.07	31.57	128.01	1334.52
42.	**769.750**	**2169.83**	**1.16**	**1.00**	**11.44**	**1.14**	**105.69**	**36.24**	**137.59**	**2529.53**
43.	831.930	2038.22	1.19	0.93	11.70	1.17	91.45	35.49	139.38	1335.70

See Table A.1 footer

Table A.6 Anthropometric analysis of male (20–40) (No. of male surveyed = 30)

No.	Height (cm)	Weight (kg)	BMI = Wt (in kg.)/Ht^2 (in m)
1.	158	50	20.02
2.	153	47	20.07
3.	162	49	18.67
4.	163	49	18.44
5.	163	55	20.70
6.	146	42	19.70
7.	140	37	18.87
8.	143	45	21.85
9.	152	55	23.81
10.	168	63	23.71
11.	160	47	18.36
12.	152	44	19.04
13.	160	57	22.27
14.	170	53	18.34
15.	163	65	24.46
16.	157	52	21.10
17.	164	47	17.37
18.	160	56	21.88
19.	166	64	23.23
20.	160	50	19.53
21.	160	50	19.53
22.	168	54	19.13
23.	170	54	18.69
24.	149	35	15.77
25.	160	55	21.27
26.	146	45	19.43
27.	160	43	16.79
28.	158	49	19.63
29.	168	52	18.42
30.	155	40	16.65

See Table A.1 footer

Table A.7 Anthropometric analysis of male (41–60) (No. of male surveyed = 27)

No.	Height (cm)	Weight (kg)	BMI = Wt (in kg.)/ Ht^2 (in m)
31.	**153**	**60**	**25.64**
32.	154	55	23.19
33.	150	40	17.78
34.	**158**	**62**	**24.91**
35.	168	48	16.90
36.	160	42.5	16.60
37.	155	30	12.49
38.	170	61	21.11
39.	170	55	19.03
40.	151	30	13.16
41.	162	47	17.91
42.	160	45	17.58
43.	**154**	**60**	**25.31**
44.	163	45	16.83
45.	156	48.5	19.92
46.	168	52	18.42
47.	150	48	21.33
48.	144	45	21.70
49.	154	47	19.82
50.	152	50	21.64
51.	160.2	55	21.43
52.	147	48	22.21
53.	154	46	19.40
54.	153	39	16.66
55.	162	65	24.76
56.	**152**	**60**	**25.97**
57.	**156**	**61**	**25.10**

See Table A.1 footer

Table A.8 Anthropometric analysis of female (20–40) (No. of female surveyed = 15)

No.	Height (cm)	Weight (kg)	BMI = Wt (in kg.)/Ht^2 (in m)
1.	147	55	25.45
2.	143	45	22.01
3.	139	51	26.40
4.	144	50	24.11
5.	146	48	22.52
6.	155	55	22.89
7.	143	45	22.01
8.	148	51	23.28
9.	152	57	24.67
10.	138	45	23.63
11.	140.6	55	27.82
12.	145.8	48	22.58
13.	148.2	55	25.04
14.	145	50	23.78
15.	137	46	24.51

See Table A.1 footer

Table A.9 Anthropometric analysis of female (41–60) (No. of female surveyed = 28)

No.	Height (cm)	Weight (kg)	BMI = Wt (in kg.)/Ht^2 (in m)
16.	138	38	19.95
17.	150	40	17.78
18.	153	49	20.93
19.	141	40	20.12
20.	**153**	**60**	**25.64**
21.	145	45	21.40
22.	145	43	20.45
23.	**149**	**58**	**26.12**
24.	140	38	19.39
25.	160	45	17.58
26.	157	44	17.85
27.	147	40	18.51
28.	148	36	16.44
29.	144.1	44	21.19
30.	145.8	45	21.17
31.	143.2	45	21.94
32.	153	50	21.36
33.	**154**	**57**	**24.05**
34.	147	35	16.20
35.	154	52	21.93
36.	150	45	20.00
37.	160	35	13.67
38.	**158**	**63**	**25.30**
39.	140	40	20.41
40.	153	45	19.22
41.	143	35	17.12
42.	**154**	**60**	**25.31**
43.	136	39	21.09

See Table A.1 footer

Table A.10 Hematological analysis of males (No. of male surveyed = 57)

No.	Hb (g/dl.)	Blood Gr.	Glucose (mg/l)	S. Urea (mg/dl)	S. Creatinine (mg%)	S. Ca (mg/dl)	S. Cholesterol (mg/dl)	HDL (mg/dl)	LDL (mg/dl)	S. triglyceride (mg/dl)	S.Na (mEq/L)	S. K (mEq/L)	S. Mg (mg/L)
1.	14.0	A+	78	25.5	0.8	9.40	180	40	123	65	137.5	3.67	18.0
2.	14.3	B+	90	34.5	0.8	9.60	215	44	95.0	105	138.5	3.45	18.5
3.	14.0	B+	80	18.4	0.6	9.32	165	38	97.0	54	133.3	3.74	18.0
4.	13.5	B+	82	19.4	0.9	9.65	170	37	109	75	136.9	4.12	21.0
5.	12.9	A+	78	16.5	1.0	9.13	165	40	127	75	136.7	3.87	17.5
6.	13.9	B+	86	38.0	1.4	10.2	175	42	110	88	138.4	4.30	23.0
7.	12.9	A+	82	23.5	1.1	9.12	160	37	80.0	60	137.5	3.55	16.0
8.	12.4	O+	80	24.5	0.6	9.10	165	38	82.0	65	136.4	3.80	20.0
9.	12.5	B+	79	24.5	0.8	10.2	175	36	85.0	89	135.0	3.62	26.0
10.	11.9	A+	80	25.0	0.9	9.12	186	43	120	110	138.2	3.60	24.0
11.	11.9	B+	90	36.0	0.7	9.43	180	37	120	120	137.5	3.56	18.0
12.	13.5	O+	89	26.5	0.6	10.2	175	45	120	110	139.5	3.65	20.0
13.	12.8	B+	78	32.0	0.8	9.50	170	38	130	95	135.9	3.50	17.0
14.	11.9	AB+	73	33.0	0.6	9.65	190	47	133	82	141.0	4.12	19.0
15.	14.2	B+	90	32.0	0.9	10.12	185	38	125	90	137.9	4.12	19.0
16.	13.4	B+	86	25.0	1.2	9.43	170	37	110	80	136.8	3.67	17.0
17.	13.0	B+	90	28.0	1.3	10.2	190	38	90.0	110	138.9	3.60	17.0
18.	13.8	O+	78	28.0	0.8	10.1	184	42	110	120	140.0	4.12	18.0
19.	14.3	B+	95	35.0	1.4	10.3	190	42	110	150	139.0	3.96	23.0
20.	14.0	B+	86	23.0	0.8	9.32	170	39	110	125	137.8	3.97	17.0
21.	11.9	A+	80	27.0	0.8	10.2	198	43	90	130	139.0	3.92	19.0
22.	12.7	A+	75	33.0	0.7	9.96	167	46	99.2	110	147.5	3.98	22.0
23.	10.9	O+	65	35.0	1.0	10.3	183	47	126	103	136.4	3.99	21.9
24.	11.5	A+	72	45.3	0.9	8.83	210	45	113	265	143.5	4.68	23.8
25.	13.7	A+	89	34.0	0.8	10.8	167	41	123	62	139. 7	3.98	22.0

(continued)

Table A.10 (continued)

No.	Hb (g/dl.)	Blood Gr.	Glucose (mg/l)	S. Urea (mg/dl)	S. Creatinine (mg%)	S. Ca (mg/dl)	S. Cholesterol (mg/dl)	HDL (mg/dl)	LDL (mg/dl)	S. triglyceride (mg/dl)	S.Na (mEq/L)	S. K (mEq/L)	S. Mg (mg/L)
26.	9.80	O+	87	34.0	0.8	9.90	216	18	99	42	142.6	4.71	18.5
27.	9.40	B+	79	34.0	0.8	10.5	193	45	113	57	137.4	4.39	22.5
28.	11.0	AB+	78	41.0	0.9	9.80	168	48	109	71	140.6	4.35	22.0
29.	11.4	B+	69	26.8	0.8	9.50	191	36	139	72	136.4	4.24	20.0
30.	12.4	B+	65	26.9	0.7	10.2	201	43	145	82	145.8	4.44	21.5
31.	12.8	B+	92	33.0	0.6	10.2	172	38	112	74	146.9	4.59	22.0
32.	10.2	AB+	92	34.0	0.9	10.4	183	43	112	117	139.9	4.29	23.0
33.	13.0	A+	80	34.0	0.7	10.4	172	39	94	148	139.0	4.40	24.0
34.	12.2	A+	77	22.5	0.7	9.30	190	41	98	100	130.5	3.59	20.8
35.	12.8	A+	67.8	28.6	0.8	8.90	180	39	102	96	129.9	3.50	21.0
36.	12.2	A+	81	18.9	0.7	8.60	170	39	99	76	131.7	3.90	23.2
37.	12.2	O+	93	26.2	0.8	10.4	160	43	106	86	135.2	3.7.0	20.0
38.	14	A+	80	30.4	0.9	9.42	195	43	112	115.5	140.3	4.42	22.4
39.	13.5	O+	70	26.7	0.7	9.20	162	42	106	59	134.8	3.76	23..0
40.	13	B+	80	21.5	0.7	9.50	170	40	116	76	140.2	3.75	20.0
41.	12	A+	76	22.5	0.9	9.13	175	39	85	87	139.5	4.10	21.0
42.	**13.5**	**AB+**	**165**	**32.5**	**1.2**	**10.2**	**180**	**41**	**160**	**157**	**138.5**	**3.90**	**18.0**
43.	11	A+	76	27.0	0.7	9.30	160	36	115	90	136.0	3.65	19.0
44.	12.9	A+	85	21.0	0.8	9.65	167	36	110	130	142.3	3.82	18.0
45.	13.9	B+	82	23.0	0.9	9.50	175	39	113	145	136.8	3.75	22.0
46.	13	B+	75	45.0	0.9	9.95	234	49	163	111	141.6	4.19	22.1
47.	**13**	**O+**	**160**	**35**	**1.1**	**9.40**	**162**	**45**	**162**	**162**	**136.4**	**3.96**	**23.0**
48.	12	O+	78.5	28.0	0.7	9.50	171	46	105	100	136.9	3.65	21.8
49.	10.4	B+	84	30.0	0.7	9.90	172	47	110	76	137.0	3.84	16.3
50.	**9.6**	**A+**	**162**	**44.0**	**1.2**	**9.50**	**162**	**37**	**155**	**148**	**133.0**	**4.12**	**20.0**

(continued)

Table A.10 (continued)

No.	Hb (g/dl.)	Blood Gr.	Glucose (mg/l)	S. Urea (mg/dl)	S. Creatinine (mg%)	S. Ca (mg/dl)	S. Cholesterol (mg/dl)	HDL (mg/dl)	LDL (mg/dl)	S. triglyceride (mg/dl)	S.Na (mEq/L)	S. K (mEq/L)	S. Mg (mg/L)
51.	11.8	A+	69	34.0	0.9	10.3	164	42	113	64	132.7	4.32	16.0
52.	12.4	B+	89	38.5	0.7	10.5	172	48	110	82	136.9	3.69	20.8
53.	12.6	B+	77	23.0	0.7	9.70	199	53	132	153	131.4	3.89	23.2
54.	9.6	B+	112	36.8	0.9	4.79	190	45	131	100	136.9	4.21	19.0
55.	10.1	AB+	86	23.9	0.9	10.3	155	44	133	131	139. 4	3.9.0	22.0
56.	**11.4**	**A+**	**166**	**38.9**	**1.3**	**9.20**	**193**	**38**	**142**	**180**	**138.9**	**4.35**	**23.8**
57.	**10.5**	**B+**	**164**	**33.8**	**1.0**	**8.98**	**174**	**44**	**148**	**172**	**139.9**	**4.49**	**22.9**

See Table A.1 footer

Table A.11 Hematological analysis of female (No. of female surveyed = 43)

No.	Hb (g/dl.)	Gr.	Glucose (mg /l)	S. Urea (mg/dl)	S. Creatinine (mg%)	S. Ca (mg/dl)	S. Cholesterol (mg/dl)	HDL (mg/dl)	LDL (mg/dl)	S. triglyceride (mg/dl)	S.Na (mEq/L)	S. K (mEq/L)	S. Mg (mg/L)
1.	10.5	O+	75.0	32.0	0.6	8.50	193	38	110	86.0	134.4	3.42	17.0
2.	12.0	A+	72.0	24.0	0.7	9.00	175	35	110	54.0	133.0	3.39	22.0
3.	13.0	B+	75.0	37.0	1.2	10.3	185	42	128	70.0	136.3	3.32	24.0
4.	13.9	B+	78.0	27.5	1.3	9.34	190	40	130	90.0	135.5	3.56	19.0
5.	10.2	B+	72.0	17.0	0.8	9.12	175	37	110	95.0	137.5	3.50	19.0
6.	11.2	B+	75.0	16.0	0.7	9.12	160	38	82.0	125	136.5	3.52	18.0
7.	10.9	B+	78.0	32.0	0.8	9.32	185	43	110	120	137.1	3.87	20.0
8.	10.0	B+	78.0	17.0	0.7	9.00	150	38	101	130	136.7	4.67	23.0
9.	11.5	O+	75.0	23.0	0.6	9.32	185	41	104	69.0	137.8	4.21	17.0
10.	10.8	A+	90.0	19.0	0.6	9.67	180	39	125	102	135.9	3.67	22.0
11.	10.5	A+	70.0	15.0	0.6	9.80	165	38	80.0	110	135.5	3.53	19.0
12.	11.5	A+	80..0	22.0	0.7	9.56	180	40	125	110	136.8	3.87	21.0
13.	12.4	B+	68.9	29.0	0.7	10.2	178	46	116	92.0	140.4	3.83	23.4
14.	9.90	O+	67.0	21.0	0.6	8.90	148	45	91.0	58.0	144.4	3.61	18.9
15.	9.00	A+	70.0	30.2	0.8	9.50	188	49	127	62.0	140.6	4.52	22.3
16.	12.1	B+	83.6	23.6	0.8	9.80	179	44	120	67.0	142.9	4.34	16.0
17.	11.2	AB+	80.2	29.0	0.8	10.0	198	40	149	55.0	139.9	3.59	18.2
18.	11.0	O+	80.0	32.6	0.9	10.3	123	49	60.0	82.0	141.8	4.39	21.2
19.	11.9	A+	74.0	32.0	0.7	9.80	179	40	121	55.0	138.4	4.95	17.2
20.	10.2	A+	76.0	34.0	0.7	9.80	196	43	90.0	72.0	138.4	4.59	23.4
21.	10.6	AB+	74.o	16.9	0.7	9.60	212	49	124	187	122.4	4.00	21.9
22.	13.9	O^{+}	74.0	18.9	0.8	9.40	182	46	99.0	75.0	132.3	3.51	23.0
23.	11.2	A+	75.0	16.0	0.6	9.00	160	38	90.0	70..0	136.5	3.76	19.0
24.	**13.2**	**B+**	**162**	**37.6**	**1.5**	**10.3**	**175**	**39**	**170**	**160**	**136.5**	**3.42**	**16.9**
25.	12.5	O+	86.0	34.0	0.9	10.2	174	40	110	75.0	135.0	3.67	21.0

(continued)

Table A.11 (continued)

No.	Hb (g/dl.)	Gr.	Glucose (mg /l)	S. Urea (mg/dl)	S. Creatinine (mg%)	S. Ca (mg/dl)	S. Cholesterol (mg/dl)	HDL (mg/dl)	LDL (mg/dl)	S. triglyceride (mg/dl)	S.Na (mEq/L)	S. K (mEq/L)	S. Mg (mg/L)
26.	9.50	B+	75.0	17.0	0.8	9.00	155	35	85.0	45.0	136.7	3.49	16.0
27	10.2	A+	80	32	0.7	10.12	190	39	97.0	139	139.2	3.67	17.0
28.	**13.1**	**B+**	**160**	**35.7**	**1.7**	**10.23**	**164**	**44**	**158**	**129**	**142.0**	**3.65**	**24.0**
29.	12.0	O+	85.1	19	0.6	9.12	170	37	90.0	127	136.0	3.57	24.0
30	11.2	A+	78.0	17	0.7	9.0	160	39	110	102	135.5	3.59	21.0
31.	10.9	O+	82.0	15.5	0.5	9.0	155	37	90.0	110	137.0	3.54	24.0
32.	**12.3**	**O+**	**166**	**34.4**	**1.6**	**9.98**	**201**	**39**	**139**	**184**	**139.5**	**3.95**	**23.0**
33	9.8.0	B+	72.0	41	0.7	9.0	162	38	113	74	141.3	4.12	20.0
34.	12.4	A+	77.0	37	0.8	9.2	167	48	100	113	147.9	3.39	20.9
35.	12.4	A+	80.0	32	0.7	8.5	200	45	114	52	143.4	4.15	21.5
36.	11.6	B+	72.0	26.6	0.6	10.3	273	48	209	91	142.0	3.69	24.0
37.	**12.8**	**A+**	**162**	**31.8**	**1.1**	**9.5**	**170**	**42**	**149**	**169**	**136.8**	**3.98**	**23.1**
38.	11.4	B+	72.0	26.8	0.7	9.6	186	42	129	61	132.9	4.37	19.2
39.	10	B+	89.0	21	0.6	10.5	164	50	99.0	79	135.1	3.82	24.0
40.	**9.4**	**O+**	**163**	**39.9**	**1.8**	**9.9**	**165**	**41**	**155**	**146.7**	**142.4**	**3.89**	**18.0**
41.	11.8	O+	76.0	22	0.8	9.5	235	46	92.0	127	140.4	4.69	21.0
42.	12	O+	78.0	34	0.5	10.3	191	50	137	89	142.0	4.36	20.0
43.	9.4	A+	80.0	33.4	1.0	9.7	214	51	185	186	134.7	3.90	24.0

See Table A.1 footer

Appendix B
Standard Methodologies Followed

Nutritional Survey and Analysis

Dietary habits of a population is determined mainly by the availability of food. For sustaining the healthy and active life, diet should be planned on sound nutritional principles. Human needs a wide range of nutrients to lead a healthy and active life, and these are derived from the diet. The amount of each nutrient that is required by man depends upon his age, energy requirement and physiological status. Adults need nutrients for maintaining constant body weight and ensuring proper body function. A little disturbance in the food habit and diet may alter the body response causing fatal diseases like obesity, hyperlipidemia, heart ailments, diabetes, and so therefore it was essential to gather information about the dietary pattern and nutritional status of the selected population.

The following standard methods (Thimmayamma et al. 2003) are applied for quantitative and qualitative evaluation of diet and nutritional status:

Diet Analysis

- *Diet History*: This method is useful for obtaining qualitative details of diet and studying patterns of food consumption at household. This procedure includes assessment of the frequency of consumption of different food—daily or number of times in a week or fortnight or occasionally. This method is used to study:

  a. Meal patterns
  b. Dietary habits
  c. Peoples food preferences and avoidances

N. Kumar, *Biogeogens and Human Health*, SpringerBriefs in Public Health,
DOI: 10.1007/978-81-322-1084-9, © The Author(s) 2013

Oral Questionnaire

This is used for quantitative estimation of nutrition. In this (24 h) recall method of oral questionnaire diet survey (Standard questionnaire given below), a set of 'standardized cups' suited to local conditions are used. The steps involved are: (1) The housewife or the member of the household who invariably cooks and serves food to the family members is asked about the types of food preparations made at breakfast, lunch, afternoon tea time and dinner. (2) An account of the raw ingredients used for each of the preparations is obtained. (3) Information on the total cooked amount of each preparation is noted in terms of standardized cup(s). (4) The intake of each food item (preparation) by the specific individual in the family is assessed by using the cups. The cups are used mainly to aid the respondent recall the quantities prepared and fed to the individual members.

Nutritional assessment

1. Which type of fuel is used in cooking? Wood/Coal/Gas/Stove/Heater/Non-smoking stove/ Other/Kanda

2. Whether Vegetarian? Yes/No
If not—which type of non-vegetarian food?

Meat type	**Daily**	**Once a week**	**Twice a week**	**Once in two weeks**	**Once a month**	**Occasional**

3. Which type of vessels or pots are used for cooking: Steel/Iron/Aluminum

4. Type of Cooking medium:

Oil	**Quantity/day**	**Quantity per month**

5. Type of meal: Plain/Spicy

6. Use of butter/ghee in diet? Yes/No
   If yes ,in what quantity: One teaspoon/Two/Three

7. Type of milk utilized: Cow milk/Buffalo/Goa/Other **Quantity**

8. Use of curd/whey/matha: Yes/No
   If yes: Type used: Curd/Whey/Matha **Quantity**

9. (a) Beverages taken? Tea/Milk/Other
   (b) How many times? Once/Twice/Many times

10. (a) Quantity of sugar in day: Once/Twice/Many times
    (b) Quantity of jaggery in day (½ piece/full)

Nutritional assessment	
11. Pulse taken with meal: Yes/No, if yes:	
Type	**Quantity (pot wise)**
12. Rice taken with meal: Yes/No, if yes:	
Variety	**Quantity (pot wise)**
13. Chapattis of which flour is used: Wheat/Bajra/Sorghum/Other	
Type of cereal	**Total Quantity**
14. Use of green vegetables: Yes/No If yes	
Varieties	**Quantity/day**
15. Use of Sugar/Beet/Other sugar-rich plant parts in diet:	
16. Number of meals? Once/Twice/Thrice	
17. Water intake?	
18. Pot wise details of the diet:	

No. of diet	**First Day Food items with quantity**	**Second Day Food items with quantity**	**Third Day Food items with quantity**
Breakfast			
Lunch			
Evening teatime			
Dinner			
Before sleep			
In between Breakfast/Lunch/Evening tea time/ Dinner			

Specific Habit

1. Smoking Yes/No Cigarettes /Bidi/Cigar/ Other with quantity
2. Alcohol consumption Yes/No Pouch/Beer/Rum/Local with quantity
3. Tobacco / Ghutka chewing Yes/No Quantity
4. Tea/ Coffee with no. of cups and sugar

Medical History

1. Medical history of a person and his family.
2. Is suffering from any disease; especially diabetes?
3. Under treatment for diabetes?

Nutritional Analysis

After tabulating the qualitative and quantitative amount of food consumed by the population in the study, the nutrients derived from the food are also calculated. An average amount of daily intake of different food types is assessed and verified by repetitive and participatory observations. Again it is crosschecked by applying Weighment Method (Thimmayamma et al. 2003) applying the following formula.

$$\frac{\textit{Intake per person}}{\textit{per day}\,(g/ml)} = \frac{\text{Raw } \textit{amount of each food}\,(\text{g})}{\textit{No. of persons} \times \textit{No. of days of survey}}$$

Thus, the accuracy is obtained to a large extent in determining the exact amount of food items consumed by the individuals on daily basis. Then, the nutritious value of these are calculated as per recommendations of **ICMR** Advisory **Committee (1989**) (Swaminathan et al. 2001); and standard method prescribed by the National Institute of Nutrition **(ICMR)** (Gopalan et al. 2003) using the Food Composition Table for calculation of different amount of nutrients and their energy value.

Anthropometric Analysis

Anthropometric data (height and weight) is obtained from the adult persons of the households following the method suggested by Weiner and Lourie. According to this method, the body weight (in kg.) is taken on a spring weighing machine, asking the subject to stand on it with an erect posture with light apparel. Height is measured by the vertical distance from the floor to the vertex by the anthropometer which is placed at the back and between the heels of the subject, taking care that it is kept absolutely vertical. Reading in centimeter and its fraction is then recorded.

The weight and height ratio is expressed in terms of body mass index which is directly associated with the dietary intake and nutritional status of an adult individual. BMI is used for assessment of chronic energy deficiency (CED) as proposed by James, Ferro-Luzzi and Waterlow, (Ferro-Luzzi et al. 1992; Naidu and Rao 1994; WHO 2000). The body mass index is most widely used because its investigation is inexpensive, noninvasive, and suitable for large scale surveys Bose et al. 2005; Khongsdier 2001; Ulijaszek et al. 1999. Thus, BMI is the most established anthropometric indication used for assessment of adult nutritional status (Moy et al. 2003). BMI is defined as the individual's body weight divided by the square of the height, and is almost always expressed in the unit kg/m^2. Hence, using the units in parentheses the BMI value can be calculated as:

$$\mathrm{BMI} = \frac{\text{weight (kg)}}{\text{height} \times \text{height (m} \times \text{m)}}$$

As per WHO (WHO 2000) the corresponding BMI and status of health is expressed as under:

BMI under 18.5 (underweight); 18.5–24.9 (normal weight); 25.0–29.9 (overweight); 30.0–34.9 (obese-Class I); 35.0–39.9 (obese-Class II); 40 or greater (obese-Class III). The index is calculated for those aged 18 and over excluding pregnant women and persons less than 3 feet (0.914 m) tall or greater than 6 feet 11 inches (2.108 m). Thus, the adults are categorized on the basis of their BMI into different stages.

Hematological Analysis

A direct manifestation of nutrition is ascertained by examining the blood sample of an individual. Since blood flows throughout the body, acting as a medium for providing oxygen and other nutrients and drawing waste products back to the excretory systems for disposal, the state of the blood stream affects, or is affected by, many medical conditions. Therefore, exact physiological status is assessed through the hematological analysis; the following parameters are described:

Determination of Calcium

Principle: Calcium reacts directly with cresolphthalein complexon (CPC) reagent containing dimethyl sulfoxide and 8 hydroxyquinoline. Since magnesium also reacts with CPC, the addition of eight hydroxyquinoline virtually eliminates the interference from magnesium (Gitleman et al. 1967; Young et al. 1975; Baginski et al. 1973; Tietz et al. 1986).

Normal Value: 9–11 mg/dl

Method: OCPC method

Requirements: Test tubes, 100-ml graduated cylinder, 100-ml beaker, 10-ml graduated pipette, push button pipette (0.05 ml), stop watch, photometer.

Reagents:

(a) *Calcium Reagent* 1: It is prepared by mixing 40 mg of cresolpthalein complexon in 1.0 ml of conc. HCl, followed by 2.5 g of 8-hydroxyquinoline, 100 ml of dimethyl sulfoxide and final quantity is made up to 1 l by using glass distilled water.
(b) *Calcium Reagent* 2: It is prepared by mixing 500 mg of potassium cyanide and 40 ml of diethyl amine in 960 ml of glass distilled water.

(c) *Calcium Standard:* 10 mg/dl (5.0 mEq/l): It contains 25 mg of $CaCO_3$ in 50% of (v/v) HCl acid.
(d) *EDTA:* 4.0 g/dl

Stability: Reagent 1 and 2 are stable at room temperature for 3 months. Reagent 3 is stable at 2–8 °C and reagent 4 is stable at room temperature for several months. *Procedure:* Fresh working reagent is prepared by mixing equal quantities of regent 1 and reagent 2.

Pipette in the tubes labeled as follows:

	Test	Standard	Blank
Working reagent (ml)	6	6	6
Serum (ml)	0.05	–	–
Standard 10 mg/dl	–	0.05	–
Distilled water (ml)	–	–	0.05

Thoroughly mixed and kept at room temperature for exactly 10 min to read intensities of test and standard against blank at 575 nm (yellow filter).

Calculation:

$$Serum\,Calcium\,\text{mg/dl} = \frac{O.D.\,Test}{O.D.\,Standard} \times 10$$

Procedure: Pipette in the tubes, labeled as follows

	Test	Standard	Blank
Working calcium (ml)	1	1	1
Serum(ml)	0.01	–	–
Ca Standard 10 mg/dl	–	0.01	–
Distilled water(ml)	–	–	0.01

Mixed and keep at room temperature (25 ± 5 °C) for 10 min. Read absorbance of test and standard against blank.

Calculation:

$$Serum\,Calcium\,\text{mg/dl} = \frac{O.D.\,Test}{O.D.\,Standard} \times 10$$

Clinical Significance: Decreased serum calcium values are found in hypoparathyroidism, rickets, osteomalacia, and steatorrhea. A fall in serum calcium can occur in acute pancreatitis and in those form of renal diseases in which excessive proteinuria is observed. Increased serum calcium values are observed in hyperparathyroidism, hypervitaminosis D and multiple myeloma. The level of calcium depends on the parathyroid hormone.

Determination of Serum Sodium and Potassium

Principle: The solution under test is passed carefully, under controlled conditions as a very fine spray in the air supply to nonluminous flame. In the flame the solution evaporates and the salt dissociates to give neutral ions, which emit light of the characteristic wave length. The flame is simultaneously monitored by both the channels. Each channel consists of a detector, which views the flame through a narrow band optical filter. The photo detector out put are connected to two independent digital display, which are calibrated for direct concentration readouts. Initial calibration is done by using at least three standards of different concentrations (Young et al. 1975).

Normal Range: Na: 135–145 mEq/L; K: 3.5–5.0 mEq/L

Method: Flame photometry method

Requirements: Test tubes, dispenser or 10-ml volumetric pipette, 10 ml beaker or bulbs, 50- or 100-μl push button pipette, flame photometer, and specimen (Serum or heparinized plasma)

Standards: Mix standards are prepared by using following two stock standards:

a. *Stock standard for sodium*: 1,000 mEq/L: It is prepared by dissolving 5.85 g of analar grade sodium chloride in glass distilled water and diluted to 100 ml by using a volumetric flask.
b. *Stock standard for potassium*: 100 mEq/L: It is prepared by dissolving 0.740 g of potassium chloride in glass distilled water and diluted to 100 ml by using a volumetric flask.

Mixed working standards are prepared as follows:

(a) Sodium/Potassium: 120/2.0 mEq/L: It contains 120 mEq of sodium and 2 mEq/L of potassium per liter of distilled water. It is prepared by mixing 12 ml of stock standard 1 and 2 ml of stock standard 2 in 86 ml of glass distilled water.
(b) Sodium/Potassium: 140/4.0 mEq/L: It is prepared by mixing 14 ml of stock standard of 1 and 4 ml of stock standard 2, in 82 ml of glass distilled water.
(c) Sodium/Potassium: 160/6.0 mEq/L: It is prepared by mixing 16 ml of stock standard of 1 and 6 ml of stock standard 2, in 78 ml of glass distilled water.

Flame Photometer: A dual channel instrument capable of quick simultaneous estimation of sodium and potassium is preferred for clinical chemistry purpose. Most of the equipment are equipped with the facilities incorporated to select Ca in place of Na and lithium in place of K Simultaneous determination of two elements minimizes sample quantity, cost of operation, and operation time.

Procedure: Pipette in the tubes, labeled as follows

	Test	Standard 1	Standard 2	Standard 3
Glass distilled water (ml)	10	10	10	10
Serum plasma (ml)	0. 1	–	–	–
Std. 120/2.0 (ml)	–	0.1	–	–
Std. 140/4.0 (ml)	–	–	0. 1	
Std. 160/6.0 (ml)	–	–	–	0.1

Mixed and transferred to beakers or bulbs for the flame photometric determination.

Clinical Significance:

a. *Hyponatremia*: Low serum sodium are observed in the condition such as: severe prolonged diarrhea and vomiting, salt losing nephritis and Addison's disease.
b. *Hyperatremia:* Increased serum sodium values are observed in the conditions such as: severe dehydration, diabetes insipidus, salt poisoning, Cushing's syndrome and in certain post renal conditions leading to obstruction to the flow of urine.
c. *Hypokalemia*: It is observed in conditions such as Cushing's syndrome, renal tubular damage, metabolic alkalosis and malnutrition.
d. *Hyperkalemia*: High potassium values are observed in the conditions such as Addison's disease, renal glomerular disease, anuria and oliguria.

Determination of Hemoglobin

Principle: In solution the ferrous ions (Fe^{2+}) of the hemoglobin are oxidized to the ferric state (Fe^{3+}) by potassium ferric cyanide to form methemoglobin. In turn methemoglobin reacts with the cyanide ions (CN^-) provided by potassium cyanide to form cyanmethemoglobin, which has the absorbance at 540 nm (Raghuramulu et al. 2003).

Normal Range: 12–14.5 (Female),
12–15.5 (Male) g/dl

Method: Cyanmethemoglobin method

Reagent:

(a) *Cyanmethemoglobin Solution (Drabkin's solution)*: 0.05 gm potassium cyanide, 0.200 g potassium ferric cyanide and 0.140 gm dihydrogen potassium phosphate is dissolved in 1 L of distilled water. 1 ml of Triton X-100 is added and mixed. Stable for at least 6 months.
(b) *Hemoglobin Standard*: Lyophilized human methemoglobin (supplied by Sigma USA). Each vial is equivalent to hemoglobin concentration of 18

g/dl whole blood when reconstituted in 50 ml of Drabkin's solution, Stable for 6 months when refrigerated at 2–6 °C.

Procedure: 0.02 ml of blood using a calibrated hemoglobin pipette is transferred into a tube containing 5.0 ml of Drabkin's reagent. The pipette is rinsed several times with the reagent; diluted hemoglobin solution is allowed to stand for at least 5 min to achieve full colour development. The absorbance is measured at 530–550 nm of the unknown sample (Aunk) and that of a standard of known hemoglobin content (Astd.) against a reagent blank.

Calculation:

$$Hemoglobin\,Unknown\,(\mathrm{g/dl}) = \frac{Aunk \times Conc.\,of\,Hemoglobin\,standard(\mathrm{g/dl})}{\mathrm{Astd.}}$$

Estimation of Blood Sugar

Principle: Glucose is oxidized to gluconic acid by glucose oxidase. The hydrogen peroxide liberated is reduced by peroxidaes and the oxygen transferred to an acceptor, which is colourless in the reduced form but coloured in the oxidized form (Raghuramulu et al. 2003; Hawh's 1965; Strehler et al. 1957; Trinder et al. 1969)·

Normal Range: 60–100 F (mg/l)
Method: Glucose oxidase method

Reagents:

(a) *Protein Precipitant*

- Sodium Tungstate $Na_2WO_42H_2O$: 10 g
- Disodium Phosphate Na_2HPO_4: 10 g
- Sodium Chloride: 9 g

The above is dissolved in about 800 ml water and approximately 125 ml of HCl is added to adjust to pH 3.0. 1 g phenol is added to makeup to 1 L with water. Stable for 1 year at 25 °C.

(b) *Colour Reagent:*

- Sodium azide 0.3 g
- 4 Aminophenazone 0.1 g
- Disodium phosphate 3 g

These are dissolved in 295 ml water, then 5 ml of glucose oxidase-peroxidase mixture is added. The mixture should have at least 1.5 µU and 3.0 µU/ml of glucose oxidase and peroxidase stable for 8 week's at 4 °C.

(c) *Standard:* The stock standard contains exactly 1 g of pure glucose in 100-ml benzoic acid solution. The working standard is made by diluting 1 ml of stock standard with 49-ml benzoic acid solution. This solution is stable at 25 °C for 1 year.

Procedure: 0.1 ml of blood is pipetted into 2.9 ml or protein precipitant; mixed well and centrifuged for about 5 min. A standard curve is set up for each batch of determination. 0.1, 0.2 and 0.3 ml of the working glucose standards equivalent to 60, 120 and 180 mg/100 ml is pipetted into clean tubes. In each case, volume is made to 1 ml with protein precipitant reagent.

Similarly, 1 ml of clear supernatant is taken from the test and placed in clean tubes. For the reagent blank, 1 ml of protein precipitant is used. To all tubes, 3 ml of colour reagent is added and incubated at 37 °C for 10 min. Then the tubes are placed in cold water for 1 min and the absorbance is read at 505 nm against the reagent blank without further delay.

The absorbance of the standard is plotted graphically and glucose values of the tests is read from this. If the glucose concentration of the test is greater than 180 mg/100 ml, the colour development stage is repeated using a smaller aliquot of the supernatant, e.g. 0.2 or 0.6 ml.

Determination of Serum Urea

Principle: Urea reacts with diacetyl—monoxime in hot acidic medium and in the presence of thiosemicarbazide and ferric ions to form a pink colored compound which can be measured on a green filter (Fearon et. al. 1939; Marsh etal. 1965; Wybenga 1970, 1071).

Normal Range (Reference range):

- Birth to one year: 4–16 mg/dl (SI: 1.4–5.7 m mol/L)
- One to forty year: 7–21 mg/dl (SI: 2.5–7.5 m mol/L)
- Gradual slight increases occur over 40 years of age.
- Possible panic range: BUN > 100 mg/dl (SI: >35.7 m mol/L)

Method: Diacetyl monoxime method

Requirements: Test tubes (15 × 25 mm), 10 ml pipette, dispenser or burette, push button pipette or 0.1-ml serological pipette, measuring cylinder of 100 ml, water bath, stop watch, and photometer.

Sample Material: Serum, heparinized plasma or fluoride plasma.

Preparation of Reagents:

(a) *Reagent 1 (DMR):* It contains 0.2 g/dl diacetyl-monoxime in distilled water. The reagent is stable at room temperature (25 ± 5 °C) for 1 year.

(b) *Reagent 2 (TSC):* It contains 40 mg/dl thiosemicarbazide in distilled water. The reagent is stable at room temperature (25 ± 5 °C) for 6 months.

(c) *Reagent 3 (Acid):* It contains 60 ml of conc. sulphuric acid. 10 ml of orthophosphoric acid and 10 ml of 1 gm /dl ferric chloride in 1 l of the reagent prepared in distilled water. This reagent is stable at room temperature for one year.
(d) *Urea Nitrogen Standard:* 20 mg/dl: It contains 42.8 mg of urea in 100 ml of saturated benzoic acid. This standard is stable for one year when refrigerated.

Preparation of Working Reagent: It is prepared fresh by mixing one part of reagent 1, one part of reagent 2, and two part of reagent 3. This reagent should be prepared fresh for each batch of the determination.

Procedure: Pipetted in the tubes labeled as follows

	Test	Standard 1	Blank
Working reagent	5.0	5.0	5.0
Serum plasma (ml)	0.05	–	–
Standard 20 mg/dl	–	0.05	–
Distilled water	–	–	0. 05

The contents of the tubes is mixed thoroughly and placed in boiling water for exactly 15 min. Cooled immediately by using tap water and after 5 min the intensities of the test and standard is measured against blank at 520 nm (green filter).

Calculation:

$$Plasma\, or (Serum)\, Urea\, Nitrogen, \text{mg/dl} = \frac{O.D.\, Test}{O.D.\, Standard} \times 20$$

Clinical Significance: Elevated levels of urea are observed in pre-renal, renal and post renal conditions.

a. *Pre-renal conditions:* Diabetes mellitus, dehydration, cardiac failure, hematemesis, severe burns, high fever, etc.
b. *Renal conditions:* Diseases of kidneys
c. *Post-renal conditions:* Enlargement of prostate, stones in the urinary tract, tumor of the bladder. Decreased values have been reported in severe liver disease, protein malnutrition and pregnancy.

Determination of Cholesterol

Principle: Cholesterol reacts with hot solution of ferric perchlorate, ethyl acetate and sulphuric acid (Cholesterol reagent) and gives a lavender colour complex which is measured at 560 nm. High density of lipoproteins is obtained in the supernatant after centrifugation. The cholesterol in the HDL fraction is also estimated by this method Dimacher et al. (1977, 1980) and Warnick et al. (1985).

Normal Value: Normal values vary with diet and age.

Total Cholesterol:

- Adult: 134–230 mg/dl
- Children: Lower values are found

HDL Cholesterol:

- Male: 35–55 mg/dl
- Female: 35–75 mg/dl

Method: One step method of Wybenga and Pileggi
Sample: Serum or Plasma
Reagent:

(a) Reagent 1: Cholesterol Reagent
(b) Reagent 2: Working Cholesterol Standard
(c) Reagent 3: Precipitating Reagent

Auxillary: Normal Saline

a. **Total Cholesterol**

Procedure: 3 ml of cholesterol reagent is added to 0.2 ml of working cholesterol Standard and 0.2 ml of specimen; pipetted into 3 test tubes labeled as Blank, Standard and Test.

	Blank	Standard	Test
Reagent 1: Cholesterol reagent (ml)	3	3	3
Reagent 2: Working cholesterol standard (ml)	–	0.2	–
Serum (ml)	–	–	0.2

Mixed well and kept immediately in the boiling water bath exactly for 90 s. Then immediately cooled to room temperature under running tap water. The optical density of standard and test against blank is measured on colorimeter with a yellow green filter at 560 nm.

b. **HDL Cholesterol**

Step I
HDL Cholesterol Separation (Supernatant)

Pipette into centrifuge tube	Quantity
Sample	0.2 ml
Precipitating reagent	0.2 ml

Mixed well and kept at room temperature for 10 min. and then centrifuged at 2000 round/min for 15 min to a clear supernatant. Then to step—II.

Step II

HDL Cholesterol Estimation: 3 ml of cholesterol reagent is added to 0.2 ml of working cholesterol standard and 0.12 ml of supernatant. Pipetted into three test tubes labeled as Blank, Standard and Test.

	Blank	Standard	Test
Reagent 1: Cholesterol reagent (ml)	3	3	3
Reagent 2: Working cholesterol standard (200 mg per) (ml)	–	0.2	–
Supernatant from step 1 (ml)	–	–	0.12

Mixed well and kept immediately in the boiling water bath exactly for 90 s. Then immediately cooled to room temperature under running tap water. The optical density of standard and test is measured against blank on a colorimeter with a yellow green filter at 560 nm.

Calculation:

1.

$$Total\,Cholesterol\,(serum/plasma)\,\mathrm{mg/dl} = \frac{OD\,Test}{}OD\,Standard \times 200$$

2.

$$HDL\,Cholesterol\,(serum/plasma\,)\mathrm{mg/dl} = \frac{T}{S} \times 200$$

Where $50 = \frac{200}{8} \times 2$

Clinical Significance: High value may be found in diabetes mellitus, hypothyroidism, obstructive jaundice, nephritic syndrome, biliary cirrhosis, athertoscelerosis, etc. Low value may be found in hyperthyroidism, malnutrition, Gaucher's disease and acute hepatitis. Decrease level of HDL cholesterol leads to increased chance of coronary heart disease while increased levels of HDL cholesterol reduce these chances. Lower values HDL cholesterol and increased ratio of total cholesterol to HDL cholesterol are taken as risk for coronary heart disease.

Determination of Triglycerides

Principle: Lipoprotein lipase hydrolyses triglycerides to glycerol and free fatty acids. The glycerol formed with ATP in the presence of glycerol kinase forms glycerol 3 phosphates, which is oxidized by enzyme glycerol phosphate oxidase to form hydrogen peroxide. The hydrogen peroxide further reacts with phenolic compound and 4 aminoantipyrine by the catalytic action of peroxidase to form a red coloured quinoneimine dye complex. Intensity of coloured formed is directly proportional to the amount of present in the sample (Bucolo et al. 1973; Dimacher et al. 1977; Trinder et al. 1969).

$$\text{Triglycerides} \overset{\text{Lipoprotein}}{\longrightarrow} \text{Lipase Glycerol} + \text{free fatty acids}$$

$$\text{Glycerol} + \text{ATP} \overset{\text{Glycerolkinase}}{\longrightarrow} \text{Glycerol 3 phosphate} + \text{ADP}$$

$$\text{Glycerol 3 phosphate} + O_2 \overset{\text{G3POxidase}}{\longrightarrow} \text{Dihydroxyacetone phosphate} + H_2O_2$$

$$H_2O_2 + \text{4 aminoantipyrine} \overset{\text{Perxidase}}{\longrightarrow} \text{Red quinoneimine dye} + H_2O + \text{phenol}$$

Normal Range:
Triglyceride: 50–190 mg/dl
150 mg/dl and above (suspicious)
200 mg/dl and above (Elevated)
LDL: 75–150 mg/dl
Method: PAP method
Content:

	25ml	2x75ml
L1 enzyme reagent 1	25	2 × 60
L2 enzyme reagent 2	5	2 × 15
S. tri. standard (200 mg/dl)	5	5

Reagent Preparation:

Working Reagent: The content of 1 bottle of L_2 is poured into 1 bottle of L_1. This working reagent is stable for at least 6 week when stored at 2–8 °C. Upon storage the working reagent may develop a slight pink colour however this does not affect the performance of the reagent. Alternatively for flexibility as much of working reagent may be made as and when desired by mixing together 4 parts of L_1 and 1 parts L_2. Alternatively 0.8 ml of L_1 and 0.2 ml of L_2 may also be used instead of 1 ml of the working reagent directly during the assay.

Triglycerdies is reported to be stable in the sample for 5 days when stored at 2–8 °C.

Procedure: Wave length/filter: 505 nm/ green; Temp.: 37 °C /RT; Line path: 1 cm.

1 ml of working reagent mixed with 0.01 ml of distilled water is added to the triglycerides standard 0.01 ml and then 0.01 ml sample is added.

Pipetted into three test tubes labeled as Blank, Standard and Test.

	Blank (ml)	Standard (ml)	Test (ml)
Working reagent	1	1	1
Distilled water	0.01	–	–
Triglycerides standard	–	0.01	–
Sample	–	–	0.01

Mixed well and incubated at 37 °C for 5 min or at room temperature (25 °C) for 15 min. The absorbance of the standard and test sample against the blank is measured within the 60 min.

Calculation:

$$\text{Triglycerides in mg/dl} = \frac{Abs.T}{Abs.S} \times 200$$

For LDL: Total cholesterol − VLDL + HDL

Where $\text{VLDL} = \frac{\text{Triglycerides test}}{5}$

Clinical Significance: Increased levels are found in hyperlipidemias, diabetes, nephritic syndrome and hypothyroidism. Increased levels are risk factor for arteriosclerotic coronary diseases and peripheral vascular diseases. Decreased levels are found in malnutrition and hyperthyroidism.

Estimation of Creatinine

Method: Jaffe method

Normal Range: 0.5–1.5 mg%

Procedure:

2 ml of plasma is mixed with 2 ml of distilled water and the protein is precipitated by adding 2 ml of 5 % sodium tungstate and 2 ml of 2/3 N sulphuric acid. 3 ml of the filtrate is mixed with 1 ml of picric acid (0.04 M solution that is 9.16 g/l) and 1 ml of NaOH (0.75 N solution). The mixture is allowed to for about 15 min. A standard is also setup with 1.5 ml of standard creatinine solution (containing 0.01 mg/ml) and 1.5 ml distilled water.

Calculation:

$$\textit{Creatinine}\ \text{mg}\ (\%) = \frac{\text{Reading of Unknown}}{}\ \text{Reading of Standard} \times 2.0$$

For the determination of creatinine 1 ml of 0.04 M solution of picric acid is added to 3 ml of protein free filtrate as previously prepared, heated in a boiling water bath for 45 min, and the volume is made up to 4 ml. With distilled water and cooled. Then 1 ml of 0.75 N sodium hydroxide is added and the creatinine is estimated as before. The results include the amount of creatinine originally present together with that formed from the creatinine. The differences of the figure obtained before and after heating the sample are multiplied by the factor 1.16. This gives the value of creatinine.

Estimation of Magnesium

Principle: Magnesium combines with Calmagite in an alkaline medium to form a red coloured complex. Interference of calcium and proteins is eliminated by the addition of specific chelating agents and detergents. Intensity of the colour formed is directly proportional to the amount of magnesium present in the sample (Fossati et al. 1982).

$$\text{Magnesium} + \text{Calmagite} \xrightarrow[\text{Medium}]{\text{Alkaline}} \text{Red coloured complex}$$

Normal Range: 6–25 mg/L
Method: Calmagite method
Reagent:
Content:

	25 ml
L1 Buffer reagent	12.5
L2 Color reagent	12.5
Magnesium standard (2.05 mEq/L)	2 ml

Reagent Preparation:
For larger assay series a working reagent may be prepared by mixing volumes of L1 (Buffer reagent) and (L2 Colour reagent). The working reagent is stable at 2–8 °C for at least one month. Keep tightly closed.
Procedure:
Wave length/filter: 546 nm/ green; Temp.: 37 °C /RT; Line path: 1 cm.
Pipette into clean dry test tubes labeled as Blank, Standard and Test.

	Blank (ml)	Standard (ml)	Test (ml)
L_1 (Buffer reagent)	0.5	0.5	0.5
L_2 (Color reagent)	0.5	0.5	0.5
Distilled water	0.01	–	–
Magnesium standard	–	0.01	–
Sample	–	–	0.01

Mixed well and incubated at room temperature (25 °C) for 5 min. The absorbance of the Standard and Test sample against Blank is measured within 30 min.

Calculation:

$$Serum\ Magnesium, \mathrm{mg/L} \frac{Abs.Test}{Abs.Standard} \times 2$$

ABO Blood Grouping

Human red blood cell antigens can be divided into four groups A, B, AB and O depending on the presence or absence of the corresponding antigens on the red blood cells. Human red blood cells possessing A and /or B antigen will agglutinate in the presence of antibody directed toward the antigen. Agglutination of red blood cells with Anti-A, Anti-B, Anti-A, B reagents is a positive test result and indicates the presence of the corresponding antigen.

Absence of agglutination of red cells with Anti-A, Anti-B, Anti-A, B reagents is a negative test result and indicates the absence of the corresponding antigen.

Reagents: Anti-A, Anti-B, Anti-A, B (of Tulip Diagnostics P Ltd., Plot Nos. 92/96, Phase II C, Verna Ind. Est., Verna, Goa-403722, India) are ready to use reagents prepared from supernatants of mouse hybridoma cell cultures. These antibodies of immunoglobulin class IgM are a mixture of several monoclonal antibodies of the same specificity but having the capability of recognizing different epitopes of the human red blood cell antigens A and B. Each batch of reagent undergoes quality control at various stages of manufacture for its specificity, avidity and performance.

Reagent Storage and Stability:

1. The reagent is stored at 2–8 °C.
2. The shelf life of reagent is as per the expiry date mentioned on the reagent vial label.

   Sample Collection: No special preparation of the patient is required prior to sample collection by approved techniques (Daniels et al. 1995; HMSO 1994; Kohler et al. 1975; Lee et al. 1983).

Additional Material Reqiured for Slide Tests: Glass Slides (50 × 75 mm), Pasteur Pipettes, Timer, Mixing Sticks.

Slide Test Procedure:

1. One drop of reagent Anti-A or Anti-B or Anti-A, B is placed on a clean glass slide.
2. To each reagent drop, one small drop of whole blood is added.
3. Mixed well with a mixing stick uniformly over an area of approximately 2.5 cm^2.
4. The slide is rocked gently, back and forth.

5. Observation is done for agglutination macroscopically at 2 min.

Interpretation of Result:

Agglutination is a positive test result indicating the presence of A and /or B antigen. Peripheral drying or fibrin strands should not be mistaken as agglutination. No agglutination is a negative test result indicating the absence of A and /or B antigen.

About the Book

Health of the people is the most important indicator of the development of a nation. Health is a state of complete physical, mental, and social wellbeing and not merely the absence of disease or infirmity (as defined by WHO). The state of health of an individual or population depends upon complex interaction of the physical, biological, political and social domains. The environment affects the human health in a big way. People tend to be most susceptible to illness when physically or mentally stressed. Stress, energy, and immunity form a closely knit network.

Through his experimental findings, the author has brought out this intricate concept of interdependence of biotic (living) and abiotic (nonliving) factors in an ecosystem, resulting in an impact on human health, in an explicitly marvelous manner. As a result, a new word "Biogeogens" has been coined, "bio" for living (biotic), "geo" for nonliving (abiotic/geographical/climatic/environment) and "gens" for the interactive proceeds of the two. The content included herein is directly concerned with the societal health and gives a clue to many socio-psycho health problems presently not handled with care. It also defines a multidimensional approach for dealing with many psychosomatic and health problems.

N. Kumar, *Biogeogens and Human Health*, SpringerBriefs in Public Health,
DOI: 10.1007/978-81-322-1084-9, © The Author(s) 2013

Printed by Publishers' Graphics LLC
DBT131215.20.06.134